Understanding Earth and Planetary Science

Understanding Earth and Planetary Science

Dustin Cook

Callisto Reference

www.callistoreference.com

Callisto Reference,
118-35 Queens Blvd., Suite 400,
Forest Hills, NY 11375, USA

Visit us on the World Wide Web at:
www.callistoreference.com

ISBN: 978-1-64116-570-9 (Hardback)

Cataloging-in-Publication Data

Understanding earth and planetary science / Dustin Cook.
p. cm.
Includes bibliographical references and index.
ISBN 978-1-64116-570-9
1. Earth sciences. 2. Planetary science. 3. Physical sciences. 4. Solar system. I. Cook, Dustin.
QE26.3 .U53 2021
550--dc23

TABLE OF CONTENTS

PREFACE

It is with great pleasure that I present this book. It has been carefully written after numerous discussions with my peers and other practitioners of the field. I would like to take this opportunity to thank my family and friends who have been extremely supporting at every step in my life.

The branch of science which deals with the study of physical constitution of the Earth and its atmosphere is referred to as Earth science. It encompasses various fields of natural science that are related to planet Earth. It includes the study of the Earth's physical characteristics. The lithosphere, the atmosphere, the biosphere and the hydrosphere are the four main areas of study within this field. It is closely related to planetary science which studies planetary systems, moons, planets, as well as their formation. Some of the major subdisciplines within this field are planetary astronomy, planetary geology and comparative planetary science. This book is compiled in such a manner, that it will provide in-depth knowledge about the theories and concepts of Earth and planetary science. Some of the diverse topics covered herein address the varied branches that fall under these categories. For someone with an interest and eye for detail, this book covers the most significant topics in the fields of Earth and Planetary Science.

The chapters below are organized to facilitate a comprehensive understanding of the subject:

Chapter - The Earth and its Spheres

The fields of natural science which deal with the study of planet Earth are known as Earth sciences. They focus on the study of the physical constitution of the Earth including lithosphere, hydrosphere, geosphere, pedosphere and atmosphere. This is an introductory chapter which will introduce briefly all these significant aspects of Earth science.

Chapter - Geology

The earth science which deals with the rocks that constitute the solid Earth as well as the processes that change them over time is referred to as geology. Some of the major branches of geology are structural geology, hydrogeology, paleontology, petroleum geology and engineering geology. These branches have been thoroughly discussed in this chapter to provide an extensive understanding of the subject.

Chapter - Atmospheric Science

Atmospheric sciences deal with the study of the Earth's atmosphere. They also deal with the study of its processes, the effects other systems have on the atmosphere, as well as the effects which the atmosphere has on these systems. Some of the major atmospheric sciences are climatology, meteorology and hydrometeorology. This chapter discusses these atmospheric sciences in detail.

Chapter - Oceanography

Oceanography is the Earth science that deals with the study of the physical and biological aspects of oceans. The branches of oceanography include geological oceanography, chemical oceanography and physical oceanography. This chapter provides comprehensive insights into oceanography and these branches.

Chapter - Hydrology

Hydrology is a discipline that studies the distribution, quality and movement of water on Earth and other planets. It focuses on the study of water resources, water cycle and environmental watershed sustainability. The diverse applications of hydrology as well as its sub-domains such as hydrometeorology and hydrometry have been thoroughly discussed in this chapter.

Chapter - Planetary Science

Planetary science is the study of the planets and their planetary systems as well as the processes related to their formation. Some of the disciplines which are closely aligned with it are planetary geology, space physics, cosmochemistry, etc. The chapter closely examines these fields of planetary science to provide an extensive understanding of the subject.

Dustin Cook

The Earth and its Spheres

1

- The Earth
- Layers of the Earth
- Spheres of the Planet Earth
- Earth Sciences
- Geography

The fields of natural science which deal with the study of planet Earth are known as Earth sciences. They focus on the study of the physical constitution of the Earth including lithosphere, hydrosphere, geosphere, pedosphere and atmosphere. This is an introductory chapter which will introduce briefly all these significant aspects of Earth science.

The Earth

Earth is the third planet from the Sun and the fifth largest planet in the solar system in terms of size and mass. Its single most outstanding feature is that its near-surface environments are the only places in the universe known to harbour life. It is designated by the symbol ♁. Earth's name in English, the international language of astronomy, derives from Old English and Germanic words for ground and earth, and it is the only name for a planet of the solar system that does not come from Greco-Roman mythology.

Since the Copernican revolution of the 16th century, at which time the Polish astronomer Nicolaus Copernicus proposed a Sun-centred model of the universe, enlightened thinkers have regarded Earth as a planet like the others of the solar system. Concurrent sea voyages provided practical proof that Earth is a globe, just as Galileo's use of his newly invented telescope in the early 17th century soon showed various other planets to be globes as well. It was only after the dawn of the space age, however, when photographs from rockets and orbiting spacecraft first captured the dramatic curvature of Earth's horizon, that the conception of Earth as a roughly spherical planet rather than as a flat entity was verified by direct human observation. Humans first witnessed Earth as a complete orb floating in the inky blackness of space in December 1968 when Apollo 8 carried astronauts around the Moon. Robotic space probes on their way to destinations beyond Earth, such as the Galileo and the Near Earth Asteroid Rendezvous (NEAR) spacecraft in the 1990s, also looked back with their cameras to provide other unique portraits of the planet.

Viewed from another planet in the solar system, Earth would appear bright and bluish in colour. Easiest to see through a large telescope would be its atmospheric features, chiefly the swirling white cloud patterns of midlatitude and tropical storms, ranged in roughly latitudinal belts around the planet. The polar regions also would appear a brilliant white, because of the clouds above and the snow and ice below. Beneath the changing patterns of clouds would appear the much darker blue-black oceans, interrupted by occasional tawny patches of desert lands. The green landscapes that harbour most human life would not be easily seen from space. Not only do they constitute a modest fraction of the land area, which itself is less than one-third of Earth's surface, but they are often obscured by clouds. Over the course of the seasons, some changes in the storm patterns and cloud belts on Earth would be observed. Also prominent would be the growth and recession of the winter snowcap across land areas of the Northern Hemisphere.

Scientists have applied the full battery of modern instrumentation to studying Earth in ways that have not yet been possible for the other planets; thus, much more is known about its structure and composition. This detailed knowledge, in turn, provides deeper insight into the mechanisms by which planets in general cool down, by which their magnetic fields are generated, and by which the separation of lighter elements from heavier ones as planets develop their internal structure releases additional energy for geologic processes and alters crustal compositions.

Earth's surface is traditionally subdivided into seven continental masses: Africa, Antarctica, Asia, Australia, Europe, North America, and South America. These continents are surrounded by four major bodies of water: the Arctic, Atlantic, Indian, and Pacific oceans. However, it is convenient to consider separate parts of Earth in terms of concentric, roughly spherical layers. Extending from the interior outward, these are the core, the mantle, the crust (including the rocky surface), the hydrosphere (predominantly the oceans, which fill in low places in the crust), the atmosphere (itself divided into spherical zones such as the troposphere, where weather occurs, and the stratosphere, where lies the ozone layer that shields Earth's surface and its organisms against the Sun's ultraviolet rays), and the magnetosphere (an enormous region in space where Earth's magnetic field dominates the behaviour of electrically charged particles coming from the Sun).

Knowledge about these divisions is summarized in this astronomically oriented overview. The discussion complements other treatments oriented to the Earth sciences and life sciences. Earth's figure and dimensions are discussed. Its magnetic field is treated in the article geomagnetic field. The geologic and biological development of Earth, including its surface features and the processes by which they are created and modified, are discussed here.

Basic Planetary Data

The mean distance of Earth from the Sun is about 149,600,000 km (92,960,000 miles). The planet orbits the Sun in a path that is presently more nearly a circle (less eccentric) than are the orbits of all but two of the other planets, Venus and Neptune. Earth makes one revolution, or one complete orbit of the Sun, in about 365.25 days. The direction of revolution—counterclockwise as viewed down from the north—is in the same sense, or direction, as the rotation of the Sun; Earth's spin, or rotation about its axis, is also in the same sense, which is called direct or prograde. The rotation period, or length of a sidereal day—23 hours, 56 minutes, and 4 seconds—is similar to that of Mars. Jupiter and most asteroids have days less than half as long, while Mercury and Venus have days more nearly comparable to their orbital periods. The 23.44° tilt, or inclination, of Earth's axis to its orbital plane, also typical, results in greater heating and more hours of daylight in one hemisphere or the other over the course of a year and so is responsible for the cyclic change of seasons.

With an equatorial radius of 6,378 km (3,963 miles), Earth is the largest of the four inner, terrestrial (rocky) planets, but it is considerably smaller than the gas giants of the outer solar system. Earth has a single natural satellite, the Moon, which orbits the planet at a mean distance of about 384,400 km (238,900 miles). The Moon is one of the bigger natural satellites in the solar system; only the giant planets have moons comparable or larger in size. Some planetary astronomers consider the Earth-Moon system a double planet, with some similarity in that regard to the dwarf planet Pluto and its largest moon, Charon.

Earth's gravitational field is manifested as the attractive force acting on a free body at rest, causing it to accelerate in the general direction of the centre of the planet. Departures from the spherical shape and the effect of Earth's rotation cause gravity to vary with latitude over the terrestrial surface. The average gravitational acceleration at sea level is about 980 cm/sec^2 (32.2 feet/sec^2).

Earth's gravity keeps the Moon in its orbit around the planet and also generates tides in the solid body of the Moon. Such deformations are manifested in the form of slight bulges at the lunar surface, detectable only by sensitive instruments. In turn, the Moon's mass—relatively large for a natural satellite—exerts a gravitational force that causes tides on Earth. The Sun, much more distant but vastly more massive, also raises tides on Earth. The tides are most apparent as the twice-daily and daily rises and falls of the ocean water, although tidal deformations occur in the solid Earth and in the atmosphere as well. The movement of the water throughout the ocean basins as a result of the tides (as well as, to a lesser extent, the tidal distortion of the solid Earth) dissipates orbital kinetic energy as heat, producing a gradual slowing of Earth's rotation and a spiraling outward of the Moon's orbit. Currently this slowing lengthens the day by a few thousandths of a second per century, but the rate of slowing varies with time as plate tectonics and sea-level changes alter the areas covered by inland bays and shallow seas.

The Outer Shell

Earth's outermost, rigid, rocky layer is called the crust. It is composed of low-density, easily melted rocks; the continental crust is predominantly granitic rock, while composition of the oceanic crust corresponds mainly to that of basalt and gabbro. Analyses of seismic waves, generated by earthquakes within Earth's interior, show that the crust extends about 50 km (30 miles) beneath the continents but only 5–10 km (3–6 miles) beneath the ocean floors.

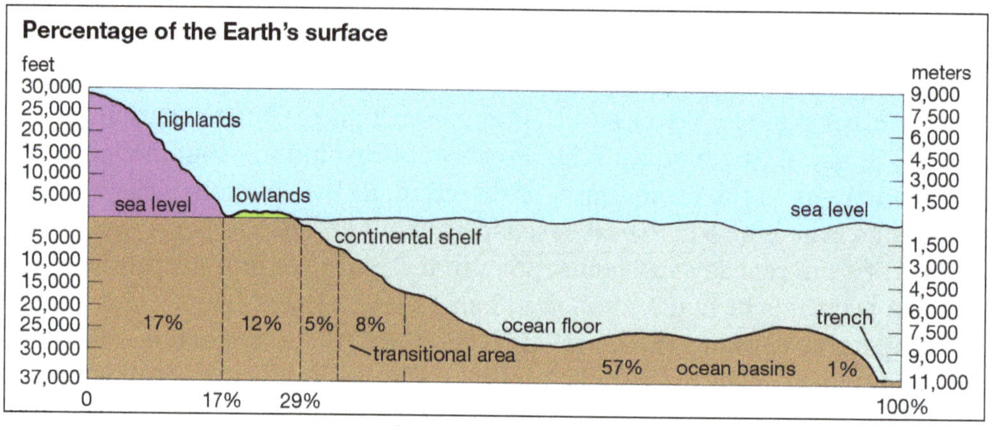

Earth: Surface composition

At the base of the crust, a sharp change in the observed behaviour of seismic waves marks the interface with the mantle. The mantle is composed of denser rocks, on which the rocks of the crust float. On geologic timescales, the mantle behaves as a very viscous fluid and responds to stress by flowing. Together the uppermost mantle and the crust act mechanically as a single rigid layer, called the lithosphere.

The lithospheric outer shell of Earth is not one continuous piece but is broken, like a slightly cracked eggshell, into about a dozen major separate rigid blocks, or plates. There are two types of plates, oceanic and continental. An example of an oceanic plate is the Pacific Plate, which extends from the East Pacific Rise to the deep-sea trenches bordering the western part of the Pacific basin. A continental plate is exemplified by the North American Plate, which includes North America as well as the oceanic crust between it and a portion of the Mid-Atlantic Ridge. The latter is an enormous submarine mountain chain that extends down the axis of the Atlantic basin, passing midway between Africa and North and South America.

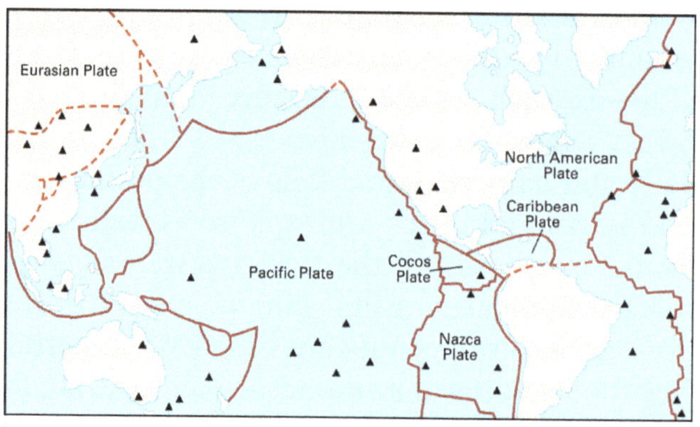

The principal tectonic plates that make up Earth's lithosphere. Also located are several dozen hot spots where plumes of hot mantle material are upwelling beneath the plates.

The lithospheric plates are about 60 km (35 miles) thick beneath the oceans and 100–200 km (60–120 miles) beneath the continents. (It should be noted that these thicknesses are defined by the mechanical rigidity of the lithospheric material. They do not correspond to the thickness of the crust, which is defined at its base by the discontinuity in seismic wave behaviour, as cited above.) They ride on a weak, perhaps partially molten, layer of the upper mantle called the asthenosphere. Slow convection currents deep within the mantle generated by radioactive heating of the interior drive lateral movements of the plates (and the continents on top of them) at a rate of several centimetres per year. The plates interact along their margins, and these boundaries are classified into three general types on the basis of the relative motions of the adjacent plates: divergent, convergent, and transform (or strike-slip).

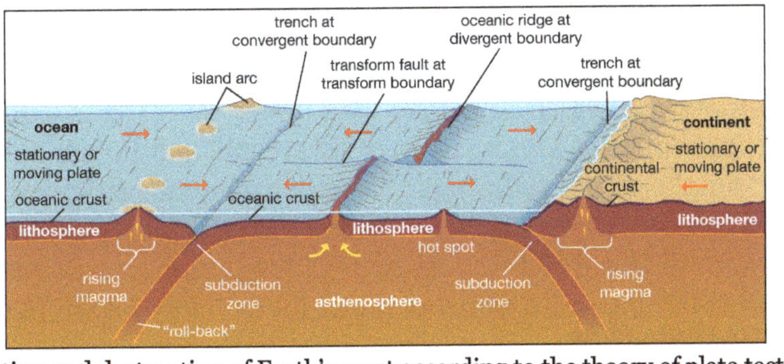

Production and destruction of Earth's crust according to the theory of plate tectonics.

Oceanic crust is continually generated at divergent plate boundaries (typified by midocean ridges and their rift zones) from upwelling mantle material, and it is consumed in the subduction process at convergent plate boundaries (marked by deep-sea trenches). Areas of convergence are sites of mountain building or of formation of volcanic island arcs. At transform, or strike-slip, boundaries, two plates slide past each other laterally; these areas are often associated with a high frequency of earthquakes.

In areas of divergence, two plates move away from each other. Buoyant upwelling motions in the mantle force the plates apart at rift zones (such as along the middle of the Atlantic Ocean floor), where magmas from the underlying mantle rise to form new oceanic crustal rocks.

Lithospheric plates move toward each other along convergent boundaries. When a continental plate and an oceanic plate come together, the leading edge of the oceanic plate is forced beneath the continental plate and down into the asthenosphere—a process called subduction. Only the thinner, denser slabs of oceanic crust will subduct, however. When two thicker, more buoyant continents come together at convergent zones, they resist subduction and tend to buckle, producing great mountain ranges. The Himalayas, along with the adjacent Plateau of Tibet, were formed during such a continent-continent collision, when India was carried into the Eurasian Plate by relative motion of the Indian-Australian Plate.

At the third type of plate boundary, the transform variety, two plates slide parallel to one another in opposite directions. These areas are often associated with high seismicity, as stresses that build up in the sliding crustal slabs are released at intervals to generate earthquakes. The San Andreas Fault in California is an example of this type of boundary, which is also known as a fault or fracture zone.

Most of Earth's active tectonic processes, including nearly all earthquakes, occur near plate margins. Volcanoes form along zones of subduction, because the oceanic crust tends to be remelted as it descends into the hot mantle and then rises to the surface as lava. Chains of active, often explosive volcanoes are thus formed in such places as the western Pacific and the west coasts of the Americas. Older mountain ranges, eroded by weathering and runoff, mark zones of earlier plate-margin activity. The oldest, most geologically stable parts of Earth are the central cores of some continents (such as Australia, parts of Africa, and northern North America). Called continental shields, they are regions where mountain building, faulting, and other tectonic processes are diminished compared with the activity that occurs at the boundaries between plates. Because of their stability, erosion has had the time to flatten the topography of continental shields. It is also on the shields that geologic evidence of crater scars from ancient impacts of asteroids and comets is better-preserved. Even there, however, tectonic processes and the action of water have erased many ancient features. In contrast, much of the oceanic crust is substantially younger (tens of millions of years old), and none dates back more than 200 million years.

This conceptual framework in which scientists now understand the evolution of Earth's lithosphere—termed plate tectonics—is almost universally accepted, although many details remain to be worked out. For example, scientists have yet to reach a general agreement as to when the original continental cores formed or how long ago modern plate-tectonic processes began to operate. Certainly the processes of internal convection, segregation of minerals by partial melting and recrystallization, and basaltic volcanism were operating more vigorously in the first billion years of Earth's history, when the planet's interior was much hotter than it is today; nevertheless, how the surface landmasses were formed and were dispersed may have been different.

Once major continental shields grew, plate tectonics was characterized by the cyclic assembly and breakup of supercontinents created by the amalgamation of many smaller continental cores and island arcs. Scientists have identified two such cycles in the geologic record. A supercontinent began breaking up about 700 million years ago, in late Precambrian time, into several major continents, but by about 250 million years ago, near the beginning of the Triassic Period, the continued drifting of these continents resulted in their fusion again into a single supercontinental landmass called Pangea. Some 70 million years later, Pangea began to fragment, gradually giving rise to today's continental configuration. The distribution is still asymmetric, with continents predominantly located in the Northern Hemisphere opposite the Pacific basin.

Startlingly, of the four terrestrial planets, only Earth shows evidence of long-term, pervasive plate tectonics. Both Venus and Mars exhibit geology dominated by basaltic volcanism on a largely immovable crust, with only faint hints of possibly limited episodes of horizontal plate motion. Mercury is intrinsically much denser than the other terrestrial planets, which implies a larger metallic core; its surface is mostly covered with impact craters, but it also shows a global pattern of scarps suggesting shrinkage of the planet, associated perhaps with interior cooling. Apparently essential to the kind of plate tectonics that occurs on Earth are large planetary size (hence, high heat flow and thin crust), which eliminates Mars, and pervasive crustal water to soften the rock, which Venus lost very early in its history. Although Earth is indeed geologically active and hence possesses a youthful surface, Venus's surface may have been completely renewed by global basaltic volcanism within the past billion years, and small portions of Mars's surface may have experienced very recent erosion from liquid water or landslides.

The Interior

More than 90 percent of Earth's mass is composed of iron, oxygen, silicon, and magnesium, elements that can form the crystalline minerals known as silicates. Nevertheless, in chemical and mineralogical composition, as in physical properties, Earth is far from homogeneous. Apart from the superficial lateral differences near the surface (i.e., in the compositions of the continental and oceanic crusts), Earth's principal differences vary with distance toward the centre. This is due to increasing temperatures and pressures and to the original segregation of materials, soon after Earth accreted from the solar nebula about 4.56 billion years ago, into a metal-rich core, a silicate-rich mantle, and the more highly refined crustal rocks. Earth is geochemically differentiated to a great extent. Crustal rocks contain several times as much of the rock-forming element aluminum as does the rest of the solid Earth and many dozens of times as much uranium. On the other hand, the crust, which accounts for a mere 0.4 percent of Earth's mass, contains less than 0.1 percent of its iron. Between 85 and 90 percent of Earth's iron is concentrated in the core.

The increasing pressure with depth causes phase changes in crustal rocks at depths between 5 and 50 km (3 and 30 miles), which marks the top of the upper mantle, as mentioned above. This transition area is called the Mohorovic ̌ić discontinuity, or Moho. Most basaltic magmas are generated in the upper mantle at depths of hundreds of kilometres. The upper mantle, which is rich in the olivine, pyroxene, and silicate perovskite minerals, shows significant lateral differences in composition. A large fraction of Earth's interior, from a depth of about 650 km (400 miles) down to 2,900 km (1,800 miles), consists of the lower mantle, which is composed chiefly of magnesium- and iron-bearing silicates, including the high-pressure equivalents of olivine and pyroxene.

The mantle is not static but rather churns slowly in convective motions, with hotter material rising up and cooler material sinking; through this process, Earth gradually loses its internal heat. In addition to being the driving force of horizontal plate motion, mantle convection is manifested in the occurrence of temporary superplumes—huge, rising jets of hot, partially molten rock—which may originate from a deep layer near the core-mantle interface. Much larger than ordinary thermal plumes, such as that associated with the Hawaiian island chain in the central Pacific, superplumes may have had profound effects on Earth's geologic history and even on its climate. One outburst of global volcanism about 66 million years ago, which created the vast flood basalt deposits known as the Deccan Traps on the Indian subcontinent, may have been associated with a superplume, though this model is far from universally accepted.

With a radius of almost 3,500 km (2,200 miles), Earth's core is about the size of the entire planet Mars. About one-third of Earth's mass is contained in the core, most of which is liquid iron alloyed with nickel and some lighter, cosmically abundant components (e.g., sulfur, oxygen, and, controversially, even hydrogen). Its liquid nature is revealed by the failure of shear-type seismic waves to penetrate the core. A small, central part of the core, however, below a depth of about 5,100 km (3,200 miles), is solid iron. This inner core is itself divided into two layers known only by the polarity differences of the iron crystals found within them. The polarity of the iron crystals of the innermost layer is oriented in an east-west direction, whereas that of the outermost layer is oriented north-south. Temperatures in the core are extremely hot, ranging from 4,000–5,000 K (roughly 6,700–8,500 °F; 3,700–4,700 °C) at the outer part of the core to 5,000–7,000 K (8,500–12,100 °F; 4,700–6,700 °C) in the centre, comparable to the surface of the Sun. Large uncertainties in temperature arise from questions as to which compounds form alloys with iron in the core, and more recent data favour the lower end of the temperature estimates for the inner core. The core's reservoir of heat may contribute as much as one-fifth of all the internal heat that ultimately flows to the surface of Earth. The basic structure of Earth—crust, mantle, and core—appears to be replicated on the other terrestrial planets, though with substantial variations in the relative size of each region.

The Geomagnetic Field and Magnetosphere

Helical fluid motions in Earth's electrically conducting liquid outer core have an electromagnetic dynamo effect, giving rise to the geomagnetic field. The planet's sizable, hot core, along with its rapid spin, probably accounts for the exceptional strength of the magnetic field of Earth compared with those of the other terrestrial planets. Venus, for example, which has a metallic core that may be similar to Earth's in size, rotates very slowly and has no detected intrinsic magnetic field. Mercury and Mars have only small intrinsic magnetic fields.

Earth's main magnetic field permeates the planet and an enormous volume of space surrounding it. A great teardrop-shaped region of space called the magnetosphere is formed by the interaction of Earth's field with the solar wind. At a distance of about 65,000 km (40,000 miles) outward toward the Sun, the pressure of the solar wind is balanced by the geomagnetic field. This serves as an obstacle to the solar wind, and the flow of charged particles, or plasma, is deflected around Earth by the resulting bow shock. The magnetosphere so produced streams out into an elongated magnetotail that stretches several million kilometres downstream from Earth away from the Sun.

Plasma particles from the solar wind can leak through the magnetopause, the sunward boundary of the magnetosphere, and populate its interior; charged particles from the Earth's ionosphere also enter the magnetosphere. The magnetotail can store for hours an enormous amount of energy—several billion megajoules, which is roughly equivalent to the yearly electricity production of many smaller countries). This occurs through a process called reconnection, in which the Sun's magnetic field, dragged into interplanetary space by the solar wind, becomes linked with the magnetic field in Earth's magnetosphere. The energy is released in dynamic structural reconfigurations of the magnetosphere, called geomagnetic substorms, which often result in the precipitation of energetic particles into the ionosphere, giving rise to fluorescing auroral displays.

Converging magnetic field lines fairly close to Earth can trap highly energetic particles so that they gyrate between the Northern and Southern hemispheres and slowly drift longitudinally around the planet in two concentric doughnut-shaped zones known as the Van Allen radiation belts. Many of the charged particles trapped in these belts are produced when energetic cosmic rays strike Earth's upper atmosphere, producing neutrons that then decay into electrons, which are negatively charged, and protons, which are positively charged. Others come from the solar wind or Earth's atmosphere. The inner radiation belt was detected in 1958 by the American physicist James Van Allen and colleagues, using a Geiger-Müller counter aboard the first U.S. satellite, Explorer 1; the outer belt was distinguished by other U.S. and Soviet spacecraft launched the same year. Earth's magnetosphere has been extensively studied ever since, and space physicists have extended their studies of plasma processes to the vicinities of comets and other planets.

An important characteristic of Earth's magnetic field is polarity reversal. In this process the direction of the dipole component reverses—i.e., the north magnetic pole becomes the south magnetic pole and vice versa. From studying the direction of magnetization of many rocks, geologists know that such reversals occur, without a discernible pattern, at intervals that range from tens of thousands of years to millions of years, though they are still uncertain about the mechanisms responsible. It is likely that during the changeover, which is believed to take a few thousand years, a nondipolar field remains, at a small fraction of the strength of the normal field. In the temporary absence of the dipole component, the solar wind would approach much closer to Earth, allowing particles that are normally deflected by the field or are trapped in its outer portions to reach the surface. The increase in particle radiation could lead to increased rates of genetic damage and thus of mutations or sterility in plants and animals, leading to the disappearance of some species. Scientists have looked for evidence of such changes in the fossil record at times of past field reversals, but the results have been inconclusive.

Development of Earth's Structure and Composition

The origin of Earth in its present form has long been the subject of intellectual interest, but since the mid-20th century scientists have made particularly significant advances both in concepts and in measurements. Analysis of the isotopes in meteorites and, in particular, of rocks brought back from the Moon by U.S. Apollo astronauts have produced some of the major contributions. Other gains have come from geochemical research on terrestrial samples combined with the new understanding of internal processes brought on by the recognition of plate tectonics, study of the terrestrial planets as a group, and advances in numerical modeling of the physical processes that lead to planetary formation.

The starting point in tracing planetary evolution is nucleosynthesis, the formation of the chemical elements on a cosmic scale. This includes the nuclear processes by which the lightest elements—mostly hydrogen and helium—were produced at the explosive birth of the universe, 13.8 billion years ago, and the subsequent formation of the heavier elements within stars. By analogy with what astronomers presently observe to happen in regions of star formation, it is thought that the solar system began as a cloud of gas and dust comprising such preexisting elements. Under its own gravitational attraction, the cloud collapsed into a rotating disk of matter, called the solar nebula. The collapse could have been initiated by a shock wave emanating from a nearby supernova, a violently exploding star, or by random density fluctuations in the cloud itself. Once sufficiently high pressures and densities were achieved in the compacted nebular core, nuclear fusion reactions within it could begin, giving birth to a star. The outer part of the rotating disk—the matter not incorporated into the new Sun—became the raw material for the planets and other orbiting bodies of the solar system. The birth of the Sun, which makes up more than 99.9 percent of the mass of the entire solar system, is taken to be the time at which the planets started to form, approximately 4.56 billion years ago.

Accretion of the Early Earth

As the gas making up the solar nebula beyond the Sun cooled with time, mineral grains are thought to have condensed and aggregated to form the earliest meteoritic material. In addition, as is suggested by the finding of anomalous concentrations of isotopes in a few meteorites, solid material from outside the solar system, apparently existing prior to the formation of the Sun, was occasionally incorporated into these developing small bodies.

The concentrations of isotopes that decay radioactively and of isotopes that are produced by radioactive decay provide scientists the information required to determine when meteorites and the planets formed. For example, the concentrations of rubidium-87 and the strontium-87 into which it decays, or those of samarium-147 and its decay product neodymium-143, indicate that the oldest meteorites formed some 4.56 billion years ago. Other isotope studies demonstrate that Earth formed within, at most, a few tens of millions of years after the birth of the Sun.

The most abundant elements in the Sun, hydrogen and helium, are severely lacking in the inner, terrestrial planets but are still abundant constituents of the large, gaseous, outer planets such as Jupiter and Saturn. It is thought that only at the distance of Jupiter and beyond—in the colder regions of the solar system, including the zone beyond Neptune in which comets originated—could the more volatile substances, which also include water, carbon dioxide, and ammonia, condense and be retained in appreciable amounts during the formation of the planets. Nevertheless, when the relative abundances of the less volatile elements are compared for the Sun, for a class of primitive, largely chemically unaltered meteorites called CI carbonaceous chondrites (considered by many researchers the most pristine samples of original solar system material), and for the estimated composition of Earth, their values are all in close agreement. This is the basis for the chondritic model, which holds that Earth (and presumably the other terrestrial planets) was essentially built up from bodies made of such meteoritic material. This idea is corroborated by isotopic studies of rocks derived from interior regions of Earth considered to be little changed throughout the planet's history. Thus, it appears that the composition of Earth is roughly what would be expected given the observed elemental abundances in the Sun and accounting for the loss of the more volatile elements.

The dust and grains that condensed out of the cooling nebular gas aggregated gravitationally to form larger fragments of rock. The chondritic meteorites observed today are basically just such collections of grains and fragments that were compacted together into larger pieces. Through continued accretion, the smaller pieces formed boulders and asteroid-size bodies (planetesimals) and, ultimately, bodies the size of the Moon and Mars. The larger the planetesimals grew, the greater their gravitational attraction and the more effectively they swept up additional particles and rock fragments while circling the Sun. Growth slowed when most bodies were lunar- and Mars-size because they were limited in number and hence effectively isolated one from another in their orbits. As Jupiter increased to its giant size, its powerful gravity perturbed these "embryos" of the terrestrial planets, elongating their orbits and allowing their incremental growth to approximately the mass of Earth to proceed over tens of millions of years.

Stony meteorites and iron meteorites (those composed largely of iron alloyed with nickel and sulfur) both fall on Earth today, and both types are thought to have been present during the formation of the planetesimals that would accrete to become Earth. In other words, Earth seems to have accreted only after most, if not all, solid matter had already condensed. Thus, a wide range of minerals was included in the grains, the larger fragments, and even the planetesimals that were accumulated by the growing planet. Apparently, such an aggregation of dense metallic fragments and less dense rocky fragments is not very stable. Calculations based on the measured strengths of rocks indicate that the metallic fragments probably sank downward as Earth grew. Although the planet was relatively cold at this stage—less than 500 K (440 °F; 230 °C)—the rock was weak. This is an important point because it leads to the conclusion that Earth's metallic core began to form during accretion of the planet and probably before the planet had grown to one-fifth of its present volume.

Effects of Planetesimal Impacts

During its accretion, Earth is thought to have been shock-heated by the impacts of meteorite-size bodies and larger planetesimals. For a meteorite collision, the heating is concentrated near the surface where the impact occurs, which allows the heat to radiate back into space. A planetesimal, however, can penetrate sufficiently deeply on impact to produce heating well beneath the surface. In addition, the debris formed on impact can blanket the planetary surface, which helps to retain heat inside the planet. Some scientists have suggested that, in this way, Earth may have become hot enough to begin melting after growing to less than 15 percent of its final volume.

Among the planetesimals striking the forming Earth, at least one is considered to have been comparable in size to Mars. Although the details are not well understood, there is good evidence that the impact of such a large planetesimal created the Moon. Among the more persuasive indications is that the relative abundances of many trace elements in rocks from the Moon are close to the values obtained for Earth's mantle. Unless this is a fortuitous coincidence, it points to the Moon having been derived from the mantle. Computer simulations have shown that a glancing collision of a Mars-size planetary body could have been sufficient to excavate from Earth's interior the material that would form the Moon. Again, the evidence for such large collisions suggests that Earth was very effectively heated during accretion.

It is apparent, then, that many processes contributing to the early development of Earth occurred almost simultaneously, within tens to hundreds of million of years after the Sun was formed. Meteorites and Earth were formed within this time, and the Moon, which has been dated at more than four billion years in age, apparently was formed in the same time period. Simultaneously, Earth's core was accumulating and may have been completely formed during the planet's growth period. In addition to the possible accretional heating caused by planetesimal impacts, the sinking of metal to form the core released enough gravitational energy to heat the entire planet by 1,000 K (1,800 °F; 1,000 °C) or more. Thus, once core formation began, Earth's interior became sufficiently hot to convect. Although it is not known whether or in what form plate tectonics was active at the surface, it seems quite possible that the underlying mantle convection began even before the planet had grown to its final dimensions. Only later in Earth's development did radioactivity become an important heat source as well.

Planetary Differentiation

Once hot, Earth's interior could begin its chemical evolution. For example, outgassing of a fraction of volatile substances that had been trapped in small amounts within the accreting planet probably formed the earliest atmosphere. Outgassing of water to Earth's surface began before 4.3 billion years ago, a time based on analysis of ancient zircons that show the effects of alteration by liquid water. In Earth's deepest interior, chemical reactions between the mantle and the core became possible. Perhaps the most important event for Earth's surface, however, was the formation of the earliest crust by partial melting of the interior. This chemical separation by partial melting and outgassing of volatiles is termed differentiation. As the interior differentiated, less-dense liquids rose from the melt toward the surface and crystallized to form crust.

Uncertainty exists over when and how the continental crust began to grow, because the record of the first 600 million years has not been found. The oldest known rocks date to only about 4 billion years. Because these are metamorphic rocks—i.e., because they were changed by heat and pressure from preexisting crustal rocks at the time of their dated age—it can be inferred that crust was present earlier in Earth's history. In fact, two tiny grains of zircon from Australia have been dated at 4.28 billion and 4.4 billion years, but their relation to the formation of continental crust is uncertain.

Although direct evidence is not available, indirect evidence derived from the compositions of rocks indicates that continental crust formed early. Isotopic analyses suggests that the average age of the present continental crust is about 2.5 billion years. Thus, in all probability, repeated partial melting of the upper mantle formed successively more refined, continent-like crustal rocks starting before 4 billion years ago. Over the first billion years, however, much of the continental crust that was formed appears to have been reincorporated into the mantle—the isotopic data infers that on average about one-third of the continental crust was recycled every billion years. As a result, only a few fragments of crust older than 3.5 billion years remain, virtually none older than 4 billion years.

The process of partial melting and formation of crust, especially continental crust, leads to a depletion of certain elements (e.g., silicon and aluminum) from the mantle. Undepleted and thus relatively primitive regions still exist, making up about one-third to one-half of the mantle, according to the isotopic models. The distribution of depleted and undepleted regions, however, is uncertain. Although much (perhaps all) of the upper mantle appears to be depleted, it is not known whether depleted rocks also exist in the lower mantle.

What is recognized is that Earth is still differentiating into chemically distinct layers or regions. This is most evident in the processes of plate tectonics that involve ongoing production of crust at divergent plate boundaries such as the midocean ridges. As this material is cycled back down into the mantle at subduction zones and then upward again, it continues to undergo chemical processing from basaltic to andesitic and eventually to granitic (continental) composition. Thus, chemical and thermal evolution of the interior, intimately connected through mantle convection, is still vigorously in progress some 4.56 billion years after the formation of the planet.

Layers of the Earth

The Earth has layers not unlike an onion and can be dissected to understand the physical and chemical properties of each layer and its influence on the rest of the Earth. Generally speaking, Earth has 4 layers:

- The outer crust that we live on
- The plastic-like mantle
- The liquid outer core
- The solid inner core

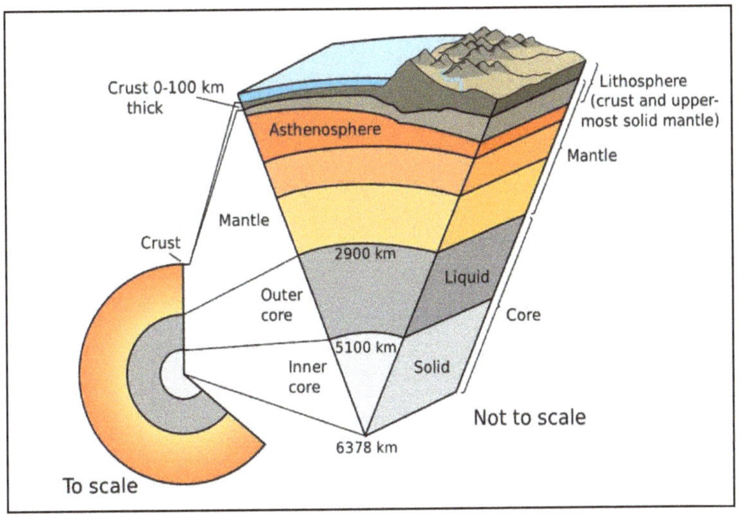

When differentiating the layers, geologists lump subdivisions into two categories, either rheologically or chemically. Rheological differentiation speaks to the liquid state of rocks under tremendous pressure and temperature. For instance, rock will respond very differently to strain under normal atmospheric temperatures and pressures as compared to fewer than thousands of kilometers of rock. If we subdivide the Earth based on rheology, we see the lithosphere, asthenosphere, mesosphere, outer core, and inner core. However, if we differentiate the layers based on chemical variations, we lump the layers into crust, mantle, outer core, and inner core.

To understand the difference in various portions of the mantle or outer versus inner core you must understand phase diagrams.

Earth's Crust

The crust is what we live on and is by far the thinnest of the layers of earth. The thickness varies depending on where you are on earth, with oceanic crust being 5-10 km and continental mountain ranges being up to 30-45 km thick. Thin oceanic crust is denser than the thicker continental crust and therefore 'floats' lower in the mantle as compared to continental crust. You will find some of the thinnest oceanic crust along mid ocean ridges where new crust is actively being formed. In comparison, when two continents collide as in the case of the India Plate and Eurasia Plate, you get some of the thickest sections of crust as it is crumpled together.

The temperatures within Earth's crust will vary from air temperatures at the surface to approximately 870 degrees Celsius in deeper sections. At this temperature, you begin to melt rock and form the below-lying mantle. Geologists subdivide Earth's crust into different plates that move about in relation to one another.

Given that Earth's surface is mostly constant in area, you cannot make crust without destroying a comparable amount of crust. With convection of the underlying mantle, we see insertion of mantle magma along mid ocean ridges, constantly forming new oceanic crust. However, to make room for this, oceanic crust must subduct (sink below) continental crust. Geologists have studied extensively the history of this plate movement, but we are sorely lacking in determining why and how these plates move the way they do.

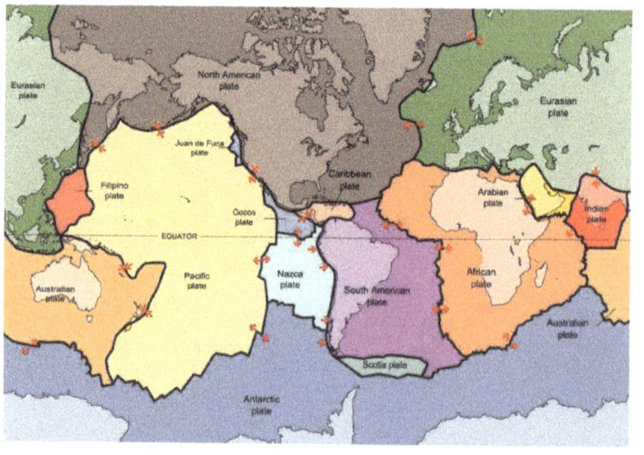

Plate tectonic map of the Earth

Earth's crust "floats" on top of the soft plastic-like mantle below. In some instances mantle clearly drives changes in the crust, as in the Hawaiian Islands. However, there is ongoing debate whether oceanic crust subduction and mid ocean ridge spreading is driven by a push or pull mechanism.

In very broad terms, oceanic crust is made up of basalt and continental crust is made up of rocks similar to granite. Below the crust is a solid relatively cooler portion of the upper mantle that is combined with the crust to make the lithosphere layer. The lithosphere is physically distinct from the below-lying layers due to its cool temperatures and typically extends 70-100 km in depth.

Below the lithosphere is the asthenosphere layer, a much hotter and malleable portion of the upper mantle. The asthenosphere begins at the bottom of the lithosphere and extends approximately 700 km into the Earth. The asthenosphere acts as the lubricating layer below the lithosphere that allows the lithosphere to move over the Earth's surface.

Earth's Mantle

The mantle is the layer of the earth that lies below the crust and is by far the largest layer making up 84% of Earth's volume. The mantle starts at the Mohorovicic Discontinuity, also known as the Moho. The Moho is defined as the density contrast from less dense crust to denser mantle and where seismic wave velocities increase. The mantle acts similar to plastic and at very high temperatures and pressures the rock is deformable at geologic timescales. This deformation causes a convection like process in the mantle where you have larige-scale upwelling and downwelling zones.

The mantle extends down to 2,890 km into the Earth's surface Temperatures that range from 500 to 900 degrees Celsius in the upper portion to over 4,000 degrees Celsius near the core boundary. Earth's mantle is believed to be composed of bulk mineralogy similar to peridotite. Gem quality peridotite is called peridot, so next time you're in a jewelry shop take a look at the peridot and you'll be looking at something similar to 84% of Earth.

Earth's Outer Core

The outer core is the liquid largely iron layer of the earth that lies below the mantle. Geologists have confirmed that the outer core is liquid due to seismic surveys of Earth's interior. The outer core is 2,300 km thick and goes down to approximately 3,400 km into the earth. No one has ever seen the outer core but based on a number of indicators, geologists believe the outer core is 80% iron, some nickel and a number of different lighter elements. When Earth was just beginning to cool billions of years ago, heavier elements sunk down into the center of the Earth, while less dense elements rose to the surface. Therefore, we see a general increase in density, as you get closer to the center of the Earth.

The outer core is hot enough to be melted but not under quite enough pressure to make the iron solid again, as seen in the inner core. The temperature of the outer core ranges from about 4,030 to 5,730 degrees Celsius. Amazingly, the outer core is fluid enough and low enough in viscosity that it may spins faster than the entire Earth. This differential velocity of spinning, along with convection and turbulent flow of the iron outer core, creates Earth's magnetic field.

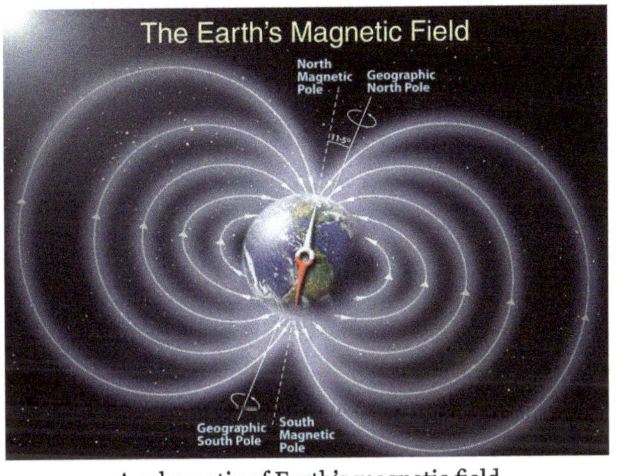

A schematic of Earth's magnetic field

Earth's Inner Core

The inner core is the centermost layer of Earth and is in many ways similar to the outer core. It is also primarily iron and nickel and has a radius of about 1,220 km. The differentiation between the outer core and inner core is density driven. The pressures become high enough that despite very high temperatures, the inner core is solid. It is also enriched in unusual heavy elements including gold, silver, platinum, palladium, and tungsten.

Temperatures reach up to 5,400 degrees Celsius and pressures up to 360 gigapascal. The inner core is about 70% of the Moon's radius and is approximately the same temperature as the surface of the Sun.

Spheres of the Planet Earth

Atmosphere

Atmosphere is the the gas and aerosol envelope that extends from the ocean, land, and ice-covered surface of a planet outward into space. The density of the atmosphere decreases outward, because the gravitational attraction of the planet, which pulls the gases and aerosols (microscopic suspended particles of dust, soot, smoke, or chemicals) inward, is greatest close to the surface. Atmospheres of some planetary bodies, such as Mercury, are almost nonexistent, as the primordial atmosphere has escaped the relatively low gravitational attraction of the planet and has been released into space. Other planets, such as Venus, Earth, Mars, and the giant outer planets of the solar system, have retained an atmosphere. In addition, Earth's atmosphere has been able to contain water in each of its three phases (solid, liquid, and gas), which has been essential for the development of life on the planet.

The evolution of Earth's current atmosphere is not completely understood. It is thought that the current atmosphere resulted from a gradual release of gases both from the planet's interior and from the metabolic activities of life-forms—as opposed to the primordial atmosphere, which developed by outgassing (venting) during the original formation of the planet. Current volcanic gaseous emissions include water vapour (H_2O), carbon dioxide (CO_2), sulfur dioxide (SO_2), hydrogen sulfide (H_2S), carbon monoxide (CO), chlorine (Cl), fluorine (F), and diatomic nitrogen (N_2; consisting of two atoms in a single molecule), as well as traces of other substances. Approximately 85 percent of volcanic emissions are in the form of water vapour. In contrast, carbon dioxide is about 10 percent of the effluent.

During the early evolution of the atmosphere on Earth, water must have been able to exist as a liquid, since the oceans have been present for at least three billion years. Given that solar output four billion years ago was only about 60 percent of what it is today, enhanced levels of carbon dioxide and perhaps ammonia (NH_3) must have been present in order to retard the loss of infrared radiation into space. The initial life-forms that evolved in this environment must have been anaerobic (i.e., surviving in the absence of oxygen). In addition, they must have been able to resist the biologically destructive ultraviolet radiation in sunlight, which was not absorbed by a layer of ozone as it is now.

Once organisms developed the capability for photosynthesis, oxygen was produced in large quantities. The buildup of oxygen in the atmosphere also permitted the development of the ozone layer as O_2 molecules were dissociated into monatomic oxygen (O; consisting of single oxygen atoms) and recombined with other O2 molecules to form triatomic ozone molecules (O_3). The capability for photosynthesis arose in primitive forms of plants between two and three billion years ago. Previous to the evolution of photosynthetic organisms, oxygen was produced in limited quantities as a by-product of the decomposition of water vapour by ultraviolet radiation.

The current molecular composition of Earth's atmosphere is diatomic nitrogen (N_2), 78.08 percent; diatomic oxygen (O_2), 20.95 percent; argon (A), 0.93 percent; water (H_2O), about 0 to 4 percent; and carbon dioxide (CO_2), 0.04 percent. Inert gases such as neon (Ne), helium (He), and krypton (Kr) and other constituents such as nitrogen oxides, compounds of sulfur, and compounds of ozone are found in lesser amounts.

Surface Budgets

Energy Budget

Earth's atmosphere is bounded at the bottom by water and land—that is, by the surface of Earth. Heating of this surface is accomplished by three physical processes—radiation, conduction, and convection—and the temperature at the interface of the atmosphere and surface is a result of this heating.

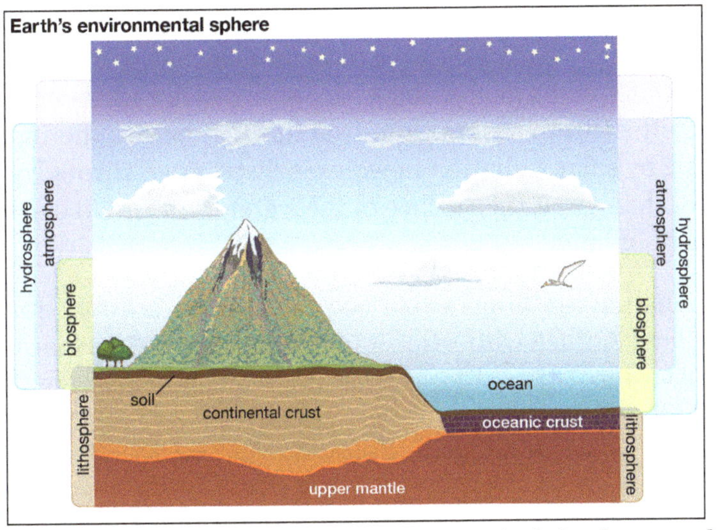

Earth's environmental spheres: Earth's environment includes the atmosphere, the hydrosphere, the lithosphere, and the biosphere.

The relative contributions of each process depend on the wind, temperature, and moisture structure in the atmosphere immediately above the surface, the intensity of solar insolation, and the physical characteristics of the surface. The temperature occurring at this interface is of critical importance in determining how suitable a location is for different forms of life.

Radiation

The temperature of the atmosphere and surface is influenced by electromagnetic radiation, and this radiation is traditionally divided into two types: insolation from the Sun and emittance from the surface and the atmosphere. Insolation is frequently referred to as shortwave radiation; it falls primarily within the ultraviolet and visible portions of the electromagnetic spectrum and consists predominantly of wavelengths of 0.39 to 0.76 micrometres (0.00002 to 0.00003 inch). Radiation emitted from Earth is called longwave radiation; it falls within the infrared portion of the spectrum and has typical wavelengths of 4 to 30 micrometres (0.0002 to 0.001 inch). Wavelengths of radiation emitted by a body depend on the temperature of the body, as specified by Planck's radiation law. The Sun, with its surface temperature of around 6,000 kelvins (K; about 5,725 °C, or 10,337 °F), emits at a much shorter wavelength than does Earth, which has lower surface and atmospheric temperatures around 250 to 300 K (−23 to 27 °C, or −9.4 to 80.6 °F).

A fraction of the incoming shortwave radiation is absorbed by atmospheric gases, including water vapour, and warms the air directly, but in the absence of clouds most of this energy reaches the surface. The scattering of a fraction of the shortwave radiation—particularly of the shortest wavelengths by air molecules in a process called Rayleigh scattering—produces Earth's blue skies.

When tall thick clouds are present, a large percentage (up to about 80 percent) of the insolation is reflected back into space. (The fraction of reflected shortwave radiation is called the cloud albedo.) Of the solar radiation reaching Earth's surface, some is reflected back into the atmosphere. Values of the surface albedo range as high as 0.95 for fresh snow to 0.10 for dark, organic soils. On land, this reflection occurs entirely at the surface. In water, however, albedo depends on the angle of the Sun's rays and the depth of the water column. If the Sun's rays strike the water surface at an oblique angle, albedo may be higher than 0.85; if these rays are more direct, only a small portion, perhaps as low as 0.02, is reflected, while the rest of the insolation is scattered within the water column and absorbed. Shortwave radiation penetrates a volume of water to significant depths (up to several hundred metres) before the insolation is completely attenuated. The heating by solar radiation in water is distributed through a depth, which results in smaller temperature changes at the surface of the water than would occur with the same insolation over an equal area of land.

The amount of solar radiation reaching the surface depends on latitude, time of year, time of day, and orientation of the land surface with respect to the Sun. In the Northern Hemisphere north of 23°30′, for example, solar insolation at local noon is less on slopes facing the north than on land oriented toward the south.

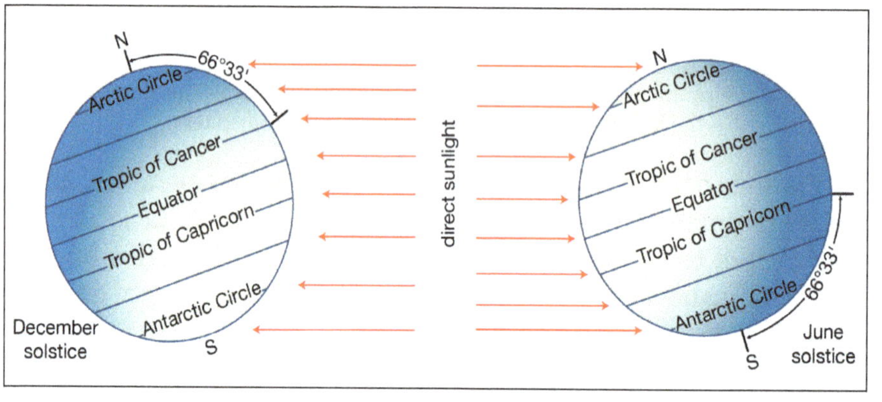

The primary cause of Earth's seasons is the change in the amount of sunlight reaching the surface at various latitudes over the course of a year. Because Earth is tilted on its axis with respect to the plane of its orbit around the Sun, different parts of its surface are in direct (overhead) sunlight at different times of the year.

Solar radiation is made up of direct and diffuse radiation. Direct shortwave radiation reaches the surface without being absorbed or scattered from its line of propagation by the intervening atmosphere. The image of the Sun's disk as a sharp and distinct object represents that portion of the solar radiation that reaches the viewer directly. Diffuse radiation, in contrast, reaches the surface after first being scattered from its line of propagation. On an overcast day, for example, the Sun's disk is not visible, and all of the shortwave radiation is diffuse.

Long-wave radiation is emitted by the atmosphere and propagates both upward and downward. According to the Stefan-Boltzmann law, the total amount of long-wave energy emitted is proportional to the fourth power of the temperature of the emitting material (e.g., the ground surface or the atmospheric layer). The magnitude of this radiation reaching the surface depends on the temperature at the height of emission and the amount of absorption that takes place between the height of emission and the surface. A larger fraction of the long-wave radiation is absorbed when the intervening atmosphere holds large amounts of water vapour and carbon dioxide. Clouds with liquid water concentrations near 2.5 grams per cubic metre absorb almost 100 percent of the long-wave radiation within a depth of 12 metres (40 feet) into the cloud. Clouds with lower liquid water concentrations require greater depths before complete absorption is attained (e.g., a cloud with a water content of 0.05 gram per cubic metre requires about 600 metres [about 2,000 feet] for complete absorption). Clouds that are at least this thick emit long-wave radiation from their bases downward to Earth's surface. The amount of long-wave radiation emitted corresponds to the temperature of the lowest levels of the cloud. (Clouds with warmer bases emit more long-wave radiation downward than colder clouds.)

Conduction

The magnitude of heat flux by conduction below a surface depends on the thermal conductivity and the vertical gradient of temperature in the material beneath the surface. Soils such as dry peat, which has very low thermal conductivity (i.e., 0.06 watt per metre per K), permit little heat flux. In contrast, concrete has a thermal conductivity about 75 times as large (i.e., 4.60 watts per metre per K) and allows substantial heat flux. In water, the thermal conductivity is relatively unimportant, since, in contrast to land surfaces, insolation extends to substantial depths in the water; in addition, water can be mixed vertically.

Convection

Vertical mixing (convection) occurs in the atmosphere as well as in bodies of water. This process of mixing is also referred to as turbulence. It is a mechanism of heat flux that occurs in the atmosphere in two forms. When the surface is substantially warmer than the overlying air, mixing will spontaneously occur in order to redistribute the heat. This process, referred to as free convection, occurs when the environmental lapse rate (the rate of change of an atmospheric variable, such as temperature or density, with increasing altitude) of temperature decreases at a rate greater than 1 °C per 100 metres (approximately 1 °F per 150 feet). This rate is called the adiabatic lapse rate (the rate of temperature change occurring within a rising or descending air parcel). In the ocean, the temperature increase with depth that results in free convection is dependent on the temperature, salinity, and depth of the water. For example, if the surface has a temperature of 20 °C (68 °F) and a salinity of 34.85 parts per thousand, an increase in temperature with depth of greater than about 0.19 °C per km (0.55 °F per mile) just below in the upper layers of the ocean will result in free convection. In the atmosphere, the temperature profile with height determines whether free convection occurs or not. In the ocean, free convection depends on the temperature and salinity profile with depth. Colder and more saline conditions in a surface parcel of water, for example, make it more likely for that parcel to sink spontaneously and thus become part of the process of free convection.

Mixing can also occur because of the shear stress of the wind on the surface. Shear stress is the pulling force of a fluid moving in one direction as it passes close to a fluid or object moving in another. As a result of surface friction, the average wind velocity at Earth's surface must be zero unless that surface is itself moving, such as in rivers or ocean currents. Winds above the surface decelerate when the vertical wind shear (the change in wind velocity at differing altitudes) becomes large enough to result in vertical mixing. The process by which heat and other atmospheric properties are mixed as a result of wind shear is called forced convection. Free and forced convection are also called convective and mechanical turbulence, respectively. This convection occurs as either sensible turbulent heat flux (heat directly transported to or from a surface) or latent turbulent heat flux (heat used to evaporate water from a surface). When this mixing does not occur, wind speeds are weak and change little with time; plumes from power-plant stacks within this layer, for example, spread very little in the vertical and remain in close proximity to the stacks.

Water Budget

The water budget at the air-surface interface is also of crucial importance in influencing atmospheric processes. The surface gains water through precipitation (rain and snow), direct condensation, and deposition (dew and frost). On land, the precipitation is often so large that some of it infiltrates into the ground or runs off into streams, rivers, lakes, and the oceans. Some of the precipitation remaining on the surface, such as in puddles or on vegetation, immediately evaporates back into the atmosphere.

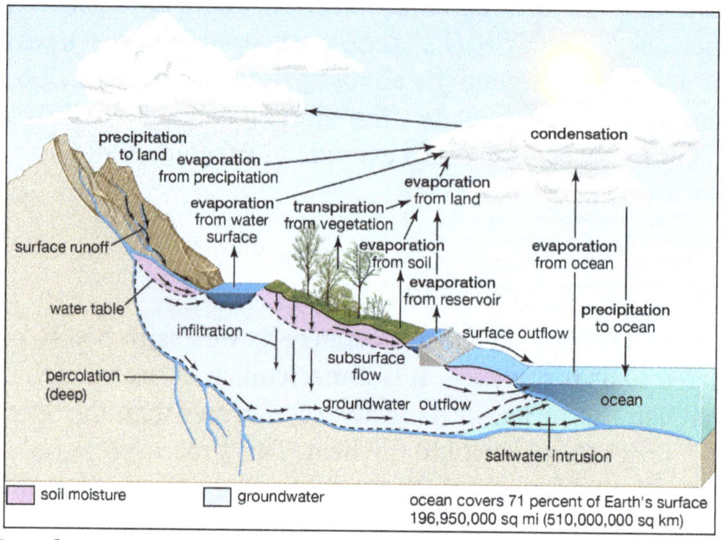

In the hydrologic cycle, water is transferred between the land surface, the ocean, and the atmosphere.

Liquid water in the soil is also converted to water vapour by transpiration from the leaves and stems of plants and by evaporation. The roots of vegetation may extract water from within the soil and emit it through stoma, or small openings, on the leaves. In addition, water may be evaporated from the surface of the soil directly, when groundwater from below is diffused upward. Evaporation occurs at the surface of water bodies at a rate that is inversely proportional to the relative humidity immediately above the surface. Evaporation is rapid in dry air but much slower when the lowest levels of the atmosphere are close to saturation. Evaporation from soils is dependent on the rate at which moisture is supplied by capillary suction within the soil, whereas transpiration from vegetation is dependent on both the water available within the root zone of plants and whether the stoma are open on the leaf surfaces. Water that evaporates and transpires into the atmosphere is often transported long distances before it precipitates out.

The input, transport, and removal of water from the atmosphere is part of the hydrologic cycle. At any one time, only a very small fraction of Earth's water is present within the atmosphere; if all the atmospheric water was condensed out, it would cover the surface of the planet only to an average of about 2.5 cm (1 inch).

Nitrogen Budget

The nitrogen budget involves the chemical transformation of diatomic nitrogen (N_2), which makes up 78 percent of the atmospheric gases, into compounds containing ammonium (NH^+), nitrite (NO_2^-), and nitrate (NO_3^-). In a process called nitrification, or nitrogen fixation, bacteria such as

Rhizobium living within nodules on the roots of peas, clover, and other legumes convert diatomic nitrogen gas to ammonia. A small amount of nitrogen is also fixed by lightning. Ammonia may be further transformed by other bacteria into nitrites and nitrates and used by plants for growth. These compounds are eventually converted back to N_2 after the plants die or are eaten by denitrifying bacteria. These bacteria, in their consumption of plants and both the excrement and corpses of plant-eating animals, convert much of the nitrogen compounds back to N_2. Some of these compounds are also converted to N_2 by a series of chemical processes associated with ultraviolet light from the Sun. The combustion of petroleum by motor vehicles also produces oxides of nitrogen, which enhance the natural concentrations of these compounds. Smog, which occurs in many urban areas, is associated with substantially higher levels of nitrogen oxides.

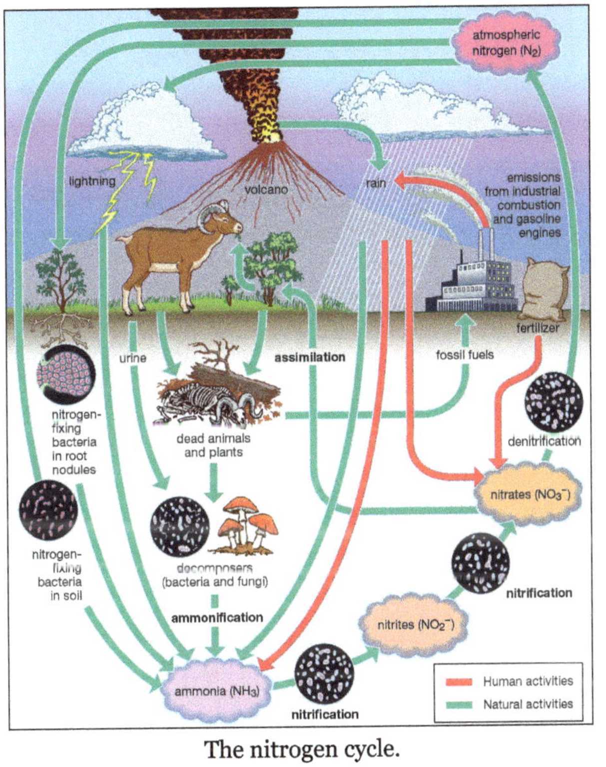

The nitrogen cycle.

Sulfur Budget

The sulfur budget is also of major importance. Sulfur is put into the atmosphere as a result of weathering of sulfur-containing rocks and by intermittent volcanic emissions. Organic forms of sulfur are incorporated into living organisms and represent an important component in both the structure and the function of proteins. Sulfur also appears in the atmosphere as the gas sulfur dioxide (SO_2) and as part of particulate compounds containing sulfate (SO_4). Alone, both are directly dry-deposited or precipitated out onto Earth's surface. When wetted, these compounds are converted to caustic sulfuric acid (H_2SO_4).

Since the beginning of the Industrial Revolution, human activities have injected significant quantities of sulfur into the atmosphere through the combustion of fossil fuels. In and near regions of urbanization and heavy industrial activity, the enhanced deposition and precipitation of sulfur in the form of sulfuric acid, and of nitrogen oxides in the form of nitric acid (HNO_3), resulting from vehicular emissions, have been associated with damage to fish populations, forests, statues, and building exteriors. The conversion of sulfur and nitrogen oxides to acids such as H_2SO_4 and HNO_3 is commonly known as the acid rain problem. Sulfur and nitrogen oxides are precipitated in rain, snow, and dry deposition (deposition to the surface during dry weather).

Mount St. Helens volcano, viewed from the south during its eruption.

Carbon Budget

The carbon budget in the atmosphere is of critical importance to climate and to life. Carbon appears in Earth's atmosphere primarily as carbon dioxide (CO_2) produced naturally by the respiration of living organisms, the decay of these organisms, the weathering of carbon-containing rock strata, and volcanic emissions. Plants utilize CO_2, water, and solar insolation to convert CO_2 to diatomic oxygen (O_2). This process, known as photosynthesis, can result in local reductions of CO_2 of tens of parts per million within vegetation canopies. In contrast, nighttime respiration occurring when photosynthesis is not active can increase CO_2 concentrations. These concentrations may even double within dense tropical forest canopies for short periods before sunrise. On the global scale, seasonal variations of about 1 percent occur as a result of CO_2 uptake from photosynthesis, plant respiration, and soil respiration. Atmospheric CO_2 is primarily absorbed in the Northern Hemisphere during the growing season (spring to autumn). CO_2 is also absorbed by ocean waters; the rate of exchange to the ocean is greater for colder than for warmer waters. Currently CO_2 makes up about 0.03 percent of the gaseous composition of the atmosphere.

Carbon is transported in various forms through the atmosphere, the hydrosphere, and geologic formations. One of the primary pathways for the exchange of carbon dioxide (CO_2) takes place between the atmosphere and the oceans; there a fraction of the CO_2 combines with water, forming carbonic acid (H_2CO_3) that subsequently loses hydrogen ions (H^+) to form bicarbonate (HCO_3^-) and carbonate (CO_{32}^-) ions. Mollusk shells or mineral precipitates that form by the reaction of calcium or other metal ions with carbonate may become buried in geologic strata and eventually release CO_2 through volcanic outgassing. Carbon dioxide also exchanges through photosynthesis in plants and through respiration in animals. Dead and decaying organic matter may ferment and release CO_2 or methane (CH_4) or may be incorporated into sedimentary rock, where it is converted to fossil fuels. Burning of hydrocarbon fuels returns CO_2 and water (H_2O) to the atmosphere. The biological and anthropogenic pathways are much faster than the geochemical pathways and, consequently, have a greater impact on the composition and temperature of the atmosphere.

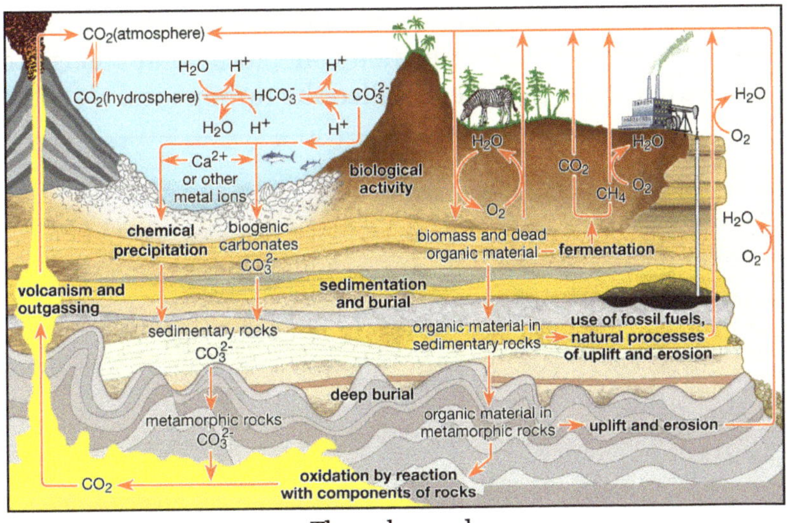

The carbon cycle

In the geologic past, CO_2 levels have been significantly higher than they are today and have had a significant effect on both climate and ecology. During the Carboniferous Period (360 to 300 million years ago), for example, moderately warm and humid climates combined with high concentrations of CO_2 were associated with extensive lush vegetation. After these plants died and decomposed, they were converted to sedimentary rocks that eventually became the coal deposits currently used for industrial combustion.

In the atmosphere, certain wavelengths of long-wave radiation are absorbed and then reemitted by CO_2. Since the lower levels of the atmosphere are warmer than layers higher up, the absorption of upward-propagating electromagnetic radiation, and a reemission of a portion of it back downward, permits the lower atmosphere to remain warmer than it would be otherwise. The association of higher concentrations of CO_2 in the air with a warmer lower troposphere is commonly referred to as the greenhouse effect. (The name is inaccurate—an actual greenhouse is warmed primarily because solar radiation enters through the glass, which retains the heated air and prevents the mixing of cooler air into the greenhouse from above.) In recent years, there has been increasing concern that the release of CO_2 through the burning of coal and other fossil fuels will warm the lower atmosphere, a phenomenon commonly referred to as global warming. Water vapour is a more efficient greenhouse gas than carbon dioxide. However, since H_2O is ubiquitous, occurring in its three phases (solid, liquid, and gas), and since CO_2 is also a biogeochemically active gas, global temperature changes are both explained and predicted by changes in the atmospheric concentration of CO_2.

Vertical Structure of the Atmosphere

Earth's atmosphere is segmented into two major zones. The homosphere is the lower of the two and the location in which turbulent mixing dominates the molecular diffusion of gases. In this region, which occurs below 100 km (about 60 miles) or so, the composition of the atmosphere tends to be independent of height. Above 100 km, in the zone called the heterosphere, various atmospheric gases are separated by molecular mass, with the lighter gases being concentrated in the highest layers. Above 1,000 km (about 600 miles), helium and hydrogen are the dominant species. Diatomic nitrogen (N_2), a relatively heavy gas, drops off rapidly with height and exists in only trace amounts at 500 km (300 miles) and above. This decrease in the concentration of heavier gases with height is largest during periods of low Sun activity, when temperatures within the heterosphere are relatively low. The transition zone, located at a height of around 100 km between the homosphere and heterosphere, is called the turbopause.

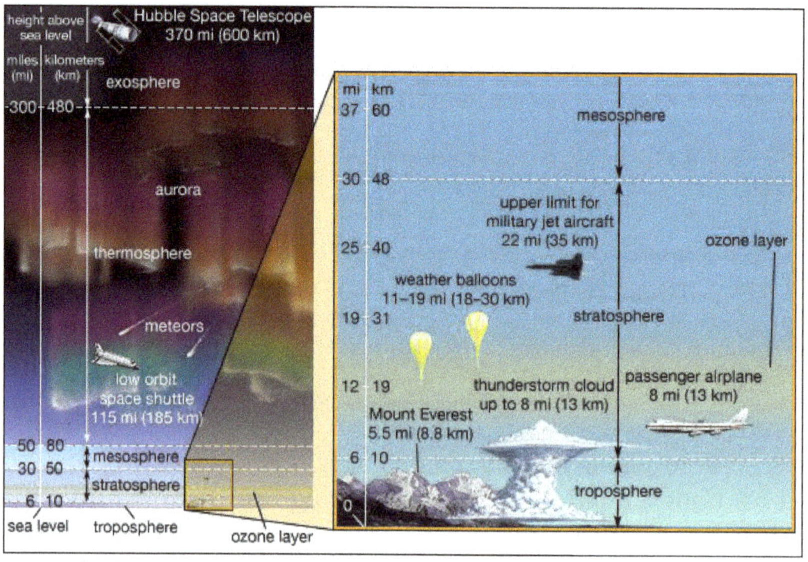

The atmosphere can be further divided into several distinct layers defined by changes in air temperature with increasing height. These layers are described below in order of increasing height above the surface.

Troposphere

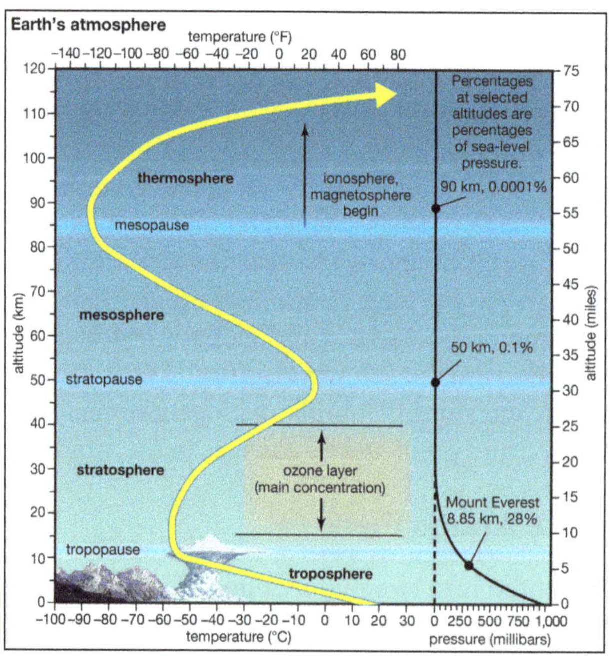

The layers of Earth's atmosphere. The yellow line shows the response of air temperature to increasing height.

The lowest portion of the atmosphere is the troposphere, a layer where temperature generally decreases with height. This layer contains most of Earth's clouds and is the location where weather primarily occurs.

Planetary Boundary Layer

The lower levels of the troposphere are usually strongly influenced by Earth's surface. This sublayer, known as the planetary boundary layer, is that region of the atmosphere in which the surface influences temperature, moisture, and wind velocity through the turbulent transfer of mass. As a result of surface friction, winds in the planetary boundary layer are usually weaker than above and tend to blow toward areas of low pressure. For this reason, the planetary boundary layer has also been called an Ekman layer, for Swedish oceanographer Vagn Walfrid Ekman, a pioneer in the study of the behaviour of wind-driven ocean currents.

Under clear, sunny skies over land, the planetary boundary layer tends to be relatively deep as a result of the heating of the ground by the Sun and the resultant generation of convective turbulence. During the summer, the planetary boundary layer can reach heights of 1 to 1.5 km (0.6 to 1 mile) above the land surface—for example, in the humid eastern United States—and up to 5 km (3 miles) in the southwestern desert. Under these conditions, when unsaturated air rises and expands, the temperature decreases at the dry adiabatic lapse rate (9.8 °C per kilometre, or roughly 23 °F per mile) throughout most of the boundary layer. Near Earth's heated surface, air temperature decreases superadiabatically (at a lapse rate greater than the dry adiabatic lapse rate). In contrast, during clear, calm nights, turbulence tends to cease, and radiational cooling (net loss of heat) from the surface results in an air temperature that increases with height above the surface.

When the rate of temperature decrease with height exceeds the adiabatic lapse rate for a region of the atmosphere, turbulence is generated. This is due to the convective overturn of the air as the warmer lower-level air rises and mixes with the cooler air aloft. In this situation, since the environmental lapse rate is greater than the adiabatic lapse rate, an ascending parcel of air remains warmer than the surrounding ambient air even though the parcel is both cooling and expanding. Evidence of this overturn is produced in the form of bubbles, or eddies, of warmer air. The larger bubbles often have sufficient buoyant energy to penetrate the top of the boundary layer. The subsequent rapid air displacement brings air from aloft into the boundary layer, thereby deepening the layer. Under these conditions of atmospheric instability, the air aloft cools according to the environmental lapse rate faster than the rising air is cooling at the adiabatic lapse rate. The air above the boundary layer replaces the rising air and undergoes compressional warming as it descends. As a result, this entrained air heats the boundary layer.

The ability of the convective bubbles to break through the top of the boundary layer depends on the environmental lapse rate aloft. The upward movement of penetrative bubbles will decrease rapidly if the parcel quickly becomes cooler than the ambient environment that surrounds it. In this situation, the air parcel will become less buoyant with additional ascent. The height that the boundary layer attains on a sunny day, therefore, is strongly influenced by the intensity of surface heating and the environmental lapse rate just above the boundary layer. The more rapidly a rising turbulent bubble cools above the boundary layer relative to the surrounding air, the lower the chance that subsequent turbulent bubbles will penetrate far above the boundary layer. The top of the daytime boundary layer is referred to as the mixed-layer inversion.

On clear, calm nights, radiational cooling results in a temperature increase with height. In this situation, known as a nocturnal inversion, turbulence is suppressed by the strong thermal stratification. Thermally stable conditions occur when warmer air overlies cooler, denser air. Over flat terrain, a nearly laminar wind flow (a pattern where winds from an upper layer easily slide past winds from a lower layer) can result. The depth of the radiationally cooled layer of air depends on a variety of factors, such as the moisture content of the air, soil and vegetation characteristics, and terrain configuration. In a desert environment, for instance, the nocturnal inversion tends to be found at greater heights than in a more humid environment. The inversion in more humid environments occurs at a lower altitude because more long-wave radiation emitted by the surface is absorbed by numerous available water molecules and reemitted back toward the surface. As a result, the lower levels of the troposphere are prevented from cooling rapidly. If the air is moist and sufficient near-surface cooling occurs, water vapour will condense into what is called "radiation fog."

Wind-generated Turbulence

During windy conditions, the mechanical production of turbulence becomes important. Turbulence eddies produced by wind shear tend to be smaller in size than the turbulence bubbles produced by the rapid convection of buoyant air. Within a few tens of metres of the surface during windy conditions, the wind speed increases dramatically with height. If the winds are sufficiently strong, the turbulence generated by wind shear can overshadow the resistance of layered, thermally stable air.

In general, there tends to be little turbulence above the boundary layer in the troposphere. Even so, there are two notable exceptions. First, turbulence is produced near jet streams, where large velocity shears exist both within and adjacent to cumuliform clouds. In these locations, buoyant turbulence occurs as a result of the release of latent heat. Second, pockets of buoyant turbulence may be found at and just above cloud tops. In these locations, the radiational cooling of the clouds destabilizes pockets of air and makes them more buoyant. Clear-air turbulence (CAT) is frequently reported when aircraft fly near one of these regions of turbulence generation.

The top of the troposphere, called the tropopause, corresponds to the level in which the pattern of decreasing temperature with height ceases. It is replaced by a layer that is essentially isothermal (of equal temperature). In the tropics and subtropics, the tropopause is high, often reaching to about 18 km (11 miles), as a result of vigorous vertical mixing of the lower atmosphere by thunderstorms. In polar regions, where such deep atmospheric turbulence is much less frequent, the tropopause is often as low as 8 km (5 miles). Temperatures at the tropopause range from as low as −80 °C (−112 °F) in the tropics to −50 °C (−58 °F) in polar regions.

Cloud Formation within the Troposphere

The region above the planetary boundary layer is commonly known as the free atmosphere. Winds at this volume are not directly retarded by surface friction. Clouds occur most frequently in this portion of the troposphere, though fog and clouds that impinge or develop over elevated terrain often occur at lower levels.

There are two basic types of clouds: cumuliform and stratiform. Both cloud types develop when clear air ascends, cooling adiabatically as it expands until either water begins to condense or deposition occurs. Water undergoes a change of state from gas to liquid under these conditions, because cooler air can hold less water vapour than warmer air. For example, air at 20 °C (68 °F) can contain almost four times as much water vapour as at 0 °C (32 °F) before saturation takes place and water vapour condenses into liquid droplets.

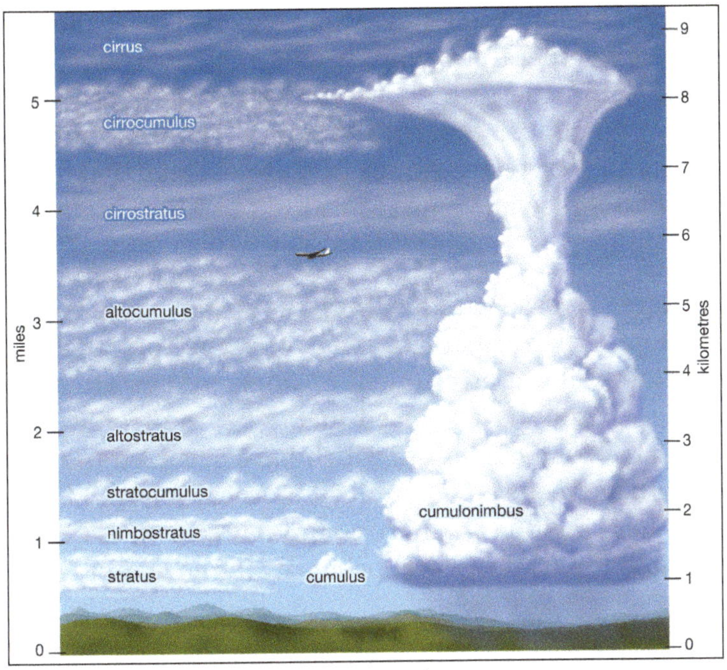

Cloud Formation

Stratiform clouds occur as saturated air is mechanically forced upward and remains colder than the surrounding clear air at the same height. In the lower troposphere, such clouds are called stratus. Advection fog is a stratus cloud with a base lying at Earth's surface. In the middle troposphere, stratiform clouds are known as altostratus. In the upper troposphere, the terms cirrostratus and cirrus are used. The cirrus cloud type refers to thin, often wispy, cirrostratus clouds. Stratiform clouds that both extend through a large fraction of the troposphere and precipitate are called nimbostratus.

Cumuliform clouds occur when saturated air is turbulent. Such clouds, with their bubbly turreted shapes, exhibit the small-scale up-and-down behaviour of air in the turbulent planetary boundary layer. Often such clouds are seen with bases at or near the top of the boundary layer as turbulent eddies generated near Earth's surface reach high enough for condensation to occur.

Cumuliform clouds will form in the free atmosphere if a parcel of air, upon saturation, is warmer than the surrounding ambient atmosphere. Since this air parcel is warmer than its surroundings, it will accelerate upward, creating the saturated turbulent bubble characteristic of a cumuliform cloud. Cumuliform clouds, which reach no higher than the lower troposphere, are known as cumulus humulus when they are randomly distributed and as stratocumulus when they are organized into lines. Cumulus congestus clouds extend into the middle troposphere, while deep, precipitating cumuliform clouds that extend throughout the troposphere are called cumulonimbus. Cumulonimbus clouds are also called thunderstorms, since they usually have lightning and thunder associated with them. Cumulonimbus clouds develop from cumulus humulus and cumulus congestus clouds.

Stratosphere and Mesosphere

The stratosphere is located above the troposphere and extends up to about 50 km (30 miles). Above the tropopause and the isothermal layer in the lower stratosphere, temperature increases with height. Temperatures as high as 0 °C (32 °F) are observed near the top of the stratosphere. The observed increase of temperature with height in the stratosphere results in strong thermodynamic stability with little turbulence and vertical mixing. The warm temperatures and very dry air result in an almost cloud-free volume. The infrequent clouds that do occur are called nacreous, or mother-of-pearl, clouds because of their striking iridescence, and they appear to be composed of both ice and supercooled water. These clouds form up to heights of 30 km (19 miles).

The pattern of temperature increase with height in the stratosphere is the result of solar heating as ultraviolet radiation in the wavelength range of 0.200 to 0.242 micrometre dissociates diatomic oxygen (O_2). The resultant attachment of single oxygen atoms to O_2 produces ozone (O_3). Natural stratospheric ozone is produced mainly in the tropical and middle latitudes. Regions of nearly complete ozone depletion, which have occurred in the Antarctic during the spring, are associated with nacreous clouds, chlorofluorocarbons (CFCs), and other pollutants from human activities. These regions are more commonly known as ozone holes. Ozone is also transported downward into the troposphere, primarily in the vicinity of the polar front.

The stratopause caps the top of the stratosphere, separating it from the mesosphere near 45–50 km (28–31 miles) in altitude and a pressure of 1 millibar (approximately equal to 0.75 mm of mercury at 0 °C, or 0.03 inch of mercury at 32 °F). In the mesosphere, temperatures again decrease with increasing altitude. Unlike the situation in the stratosphere, vertical air currents in the mesosphere are not strongly inhibited. Ice crystal clouds, called noctilucent clouds, occasionally form in the upper mesosphere. Above the mesopause, a region occurring at altitudes near 85 to 90 km (50 to 55 miles), temperature again increases with height in a layer called the thermosphere.

Thermosphere

Temperatures in the thermosphere range from near 500 K (approximately 227 °C, or 440 °F) during periods of low sunspot activity to 2,000 K (1,725 °C, or 3,137 °F) when the Sun is active. The thermopause, defined as the level of transition to a more or less isothermal temperature profile at the top of the thermosphere, occurs at heights of around 250 km (150 miles) during quiet Sun periods and almost 500 km (300 miles) when the Sun is active. Above 500 km, molecular collisions are infrequent enough that temperature is difficult to define.

The portion of the thermosphere where charged particles (ions) are abundant is called the ionosphere. These ions result from the removal of electrons from atmospheric gases by solar ultraviolet radiation. Extending from about 80 to 300 km (about 50 to 185 miles) in altitude, the ionosphere is an electrically conducting region capable of reflecting radio signals back to Earth.

Maximum ion density, a condition that makes for efficient radio transmission, occurs within two sublayers: the lower E region, which exists from 90 to 120 km (about 55 to 75 miles) in altitude; and the F region, which exists from 150 to 300 km (about 90 to 185 miles) in altitude. The F region has two maxima (i.e., two periods of highest ion density) during daylight hours, called F1 and F2. Both the F1 and F2 regions possess high ion density and are strongly influenced by both solar activity and time of day. Of these, the F2 region is the more variable of the two and may reach an ion density as high as 106 electrons per cubic centimetre. Shortwave radio transmissions, capable of reaching around the world, take advantage of the ability of layers in the ionosphere to reflect certain wavelengths of electromagnetic radiation. In addition, electrical discharges from the tops of thunderstorms into the ionosphere, called transient luminous events, have been observed.

Magnetosphere and Exosphere

Above approximately 500 km (300 miles), the motion of ions is strongly constrained by the presence of Earth's magnetic field. This region of Earth's atmosphere, called the magnetosphere, is compressed by the solar wind on the daylight side of the planet and stretched outward in a long tail on the night side. The colourful auroral displays often seen in polar latitudes are associated with bursts of high-energy particles generated by the Sun. When these particles are influenced by the magnetosphere, some are subsequently injected into the lower ionosphere.

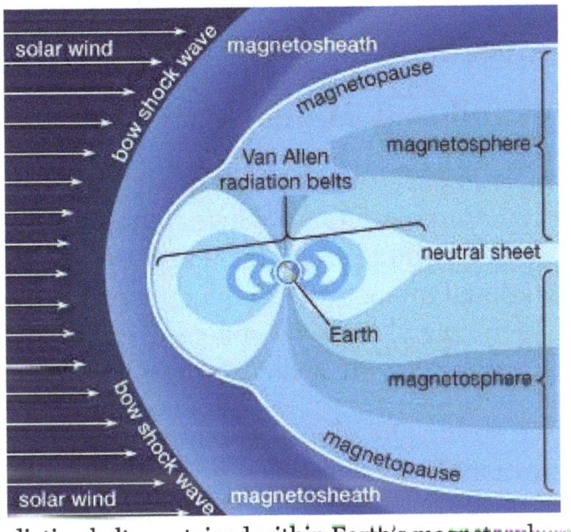

The Van Allen radiation belts contained within Earth's magnetosphere. Pressure from the solar wind is responsible for the asymmetrical shape of the magnetosphere and the belts.

The layer above 500 km is referred to as the exosphere, a region in which at least half of the upward-moving molecules do not collide with one another. In contrast, these molecules follow long ballistic trajectories and may exit the atmosphere completely if their escape velocities are high enough. The loss rate of molecules through the exosphere is critical in determining whether Earth or any other planetary body retains an atmosphere.

Horizontal Structure of the Atmosphere

Distribution of Heat from the Sun

The primary driving force for the horizontal structure of Earth's atmosphere is the amount and distribution of solar radiation that comes in contact with the planet. Earth's orbit around the Sun is an ellipse, with a perihelion (closest approach) of 147.5 million km (91.7 million miles) in early January and an aphelion (farthest distance) of 152.6 million km (94.8 million miles) in early July. As a result of Earth's elliptical orbit, the time between the autumnal equinox and the following vernal equinox (about September 22 to about March 21) is almost one week shorter than the remainder of the year in the Northern Hemisphere. This results in a shorter astronomical winter in the Northern Hemisphere than in the Southern Hemisphere.

Earth rotates once every 24 hours around an axis that is tilted at an angle of 23°30′ with respect to the plane of its orbit around the Sun. As a result of this tilt, during the summer season of either the Northern or the Southern Hemisphere, the Sun's rays are more direct at a given latitude than they are during the winter season. Poleward of latitudes 66°30′ N and 66°30′ S, the tilt of the planet is such that for at least one complete day (at 66°30′) and as long as six months (at 90°), the Sun is above the horizon during the summer season and below the horizon during the winter.

As a result of this asymmetric distribution of solar heating, during the winter season the troposphere in the high latitudes becomes very cold. In contrast, during the summer at high latitudes, the troposphere warms significantly as a result of the long hours of daylight; however, owing to the oblique angle of the sunlight near the poles, the temperatures there remain relatively cool compared with middle latitudes. Equatorward of latitudes 30° N and 30° S or so, substantial radiant heating from the Sun occurs during both winter and summer seasons. The tropical troposphere, therefore, has comparatively little variation in temperature during the year.

Convection, Circulation and Deflection of Air

The region of greatest solar heating at the surface in the humid tropics corresponds to areas of deep cumulonimbus convection. Cumulonimbus clouds routinely form in the tropics where rising parcels of air are warmer than the surrounding ambient atmosphere. They transport water vapour, sensible heat, and Earth's rotational momentum to the upper portion of the troposphere. As a result of the vigorous convective mixing of the atmosphere, the tropopause in the lower latitudes is often very high, located some 17 to 18 km (10.5 to 11 miles) above the surface.

Since motion upward into the stratosphere is inhibited by very stable thermal layering, the air transported upward by convection diverges toward the poles in the upper troposphere. (This divergence aloft results in a wide strip of low atmospheric pressure at the surface in the tropics, occurring in an area called the equatorial trough). As the diverted air in the troposphere moves toward the poles, it tends to retain the angular momentum of the near-equatorial region, which is large as a result of Earth's rotation. As a result, the poleward-moving air is deflected toward the right in the Northern Hemisphere and toward the left in the Southern Hemisphere.

The Erg Admer, a large area of sand dunes in southern Algeria, is located within a vast region of subtropical high pressure. The arid conditions here result from the constant presence of descending air containing little moisture.

Upon reaching around 30° of latitude poleward of its region of origin, the upper-level air is traveling primarily toward the poles and is tending toward the east. Since motion upward is constrained by the stratosphere, the slowly cooling air must descend. The compressional warming that occurs as the air descends creates vast regions of subtropical high pressure. These regions are centred over the oceans and are characterized by strong thermodynamic stability. The sparse precipitation in these regions, a result of stability and subsidence, is associated with such great arid regions of the world as the Sahara, Atacama, Kalahari, and Sonoran deserts. The accumulation of air as a result of the convergence in the upper troposphere causes deep high-pressure systems, known as subtropical ridges, to form in these regions. Locally, these ridges are given such names as the Bermuda High, the Azores High, and the North Pacific High.

The descending air referred to above, upon reaching the lower troposphere, is forced to diverge by the presence of Earth's surface. Some air moves poleward, while the remainder moves equatorward. In either direction, the air is deflected to the right in the Northern Hemisphere and to the left in the Southern Hemisphere. Deflection occurs because, in accordance with Newton's first law of motion, a parcel moving in a certain direction will retain the same motion unless acted on by an exterior force. With respect to a rotating Earth, a moving parcel conserving its momentum (i.e., not acted on by an exterior force) will appear to be deflected with respect to fixed points on the rotating Earth. As seen from a fixed point in space, such a parcel would be moving in a straight line. This apparent force on the motion of a fluid (in this case, air) is called the Coriolis effect. As a result of the Coriolis effect, air tends to rotate counterclockwise around large-scale low-pressure systems and clockwise around large-scale high-pressure systems in the Northern Hemisphere. In the Southern Hemisphere, the flow direction is reversed.

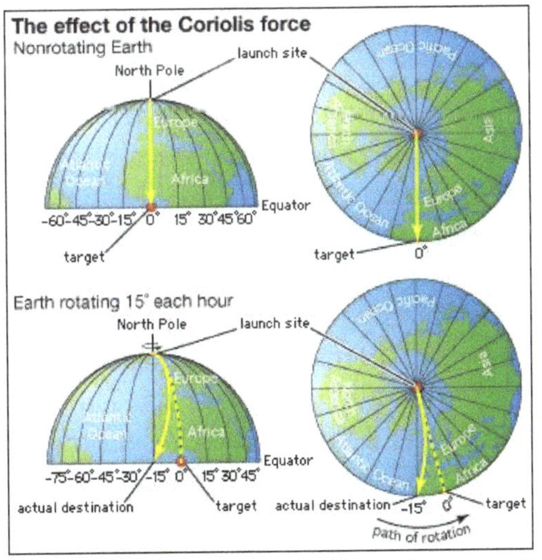

The effect of the Coriolis force (the rocket example).

In the equatorward-moving flow, this deflection results in northeast winds north of 0° latitude and southeast winds south of that latitude. These low-level winds have been called the trade winds since 17th-century sailing vessels used them to travel to the Americas. The convergence region for lower-level northeast and southeast trade winds is called the intertropical convergence zone (ITCZ). The ITCZ corresponds to the equatorial trough and is the mechanism that helps generate the deep cumulonimbus clouds through convection. Cumulonimbus clouds are the main conduit transporting tropical heating into the upper troposphere.

The circulation pattern described above—ascent in the equatorial trough, poleward movement in the upper troposphere, descent in the subtropical ridges, and equatorward movement in the trade winds—is in effect a direct heat engine, which meteorologists call the Hadley cell. This persistent circulation mechanism transports heat from the latitudes of greatest solar insolation to the latitudes of the subtropical ridges. The geographic location of the Hadley circulation moves north and south with the seasons; however, the equatorial trough lags behind for about two months owing to the thermal inertia of Earth's surface. (For a given location on Earth's surface, the highest daily temperatures are achieved just after the period of greatest insolation, since time is required to heat the ocean surface waters and the soil.)

Extratropical Cyclones

Poleward of the subtropical ridges, winds in the lower troposphere tend to be southwesterly in the Northern Hemisphere and northwesterly in the Southern Hemisphere, again owing to the Coriolis effect. Since warm air is being moved poleward at low altitudes, the wind flow is no longer associated with the direct heat engine of the Hadley cell. Instead, the continued transport of heat from the equatorial trough toward the poles is facilitated by large low-pressure eddies called extratropical cyclones. These phenomena develop along the polar front, which separates colder polar air from warmer tropical air, when sufficiently large temperature differences occur across the frontal boundary in the lower troposphere. The intensity of this temperature gradient is referred to as the baroclinicity of the front.

Extratropical cyclones have three stages of expansion: the developing stage, in which an undulating wave develops along the front; the mature stage, in which sinking cold air sweeps equatorward west of the surface low-pressure centre and ascending warm air moves poleward east of the cyclone; and the occluded stage, in which the warm air is entrained within and moved above the polar air and becomes separated from the source region of the tropical air. Cyclones that progress no farther than the developing stage are referred to as wave cyclones, while extratropical lows that reach the mature and occluded stages are called baroclinically unstable waves. Extratropical storm development is referred to as cyclogenesis. Rapid extratropical cyclone development, called explosive cyclogenesis, is often associated with major winter storms and occurs when surface pressure falls by more than about 24 millibars per day. Theoretical analysis has shown that the occurrence of baroclinically unstable waves is directly proportional to the magnitude of the temperature gradient, with maximum growth for wavelengths of 3,000 to 5,000 km (1,865 to 3,100 miles). Wavelengths that are shorter are damped by horizontal mixing. The 3,000 to 5,000 km wavelength is the typical separation between high- and low- pressure synoptic weather systems in the middle and higher latitudes.

Polar Fronts and the Jet Stream

In the troposphere, the demarcation between polar air and warmer tropical atmosphere is usually defined by the polar front. On the poleward side of the front, the air is cold and more dense; equatorward of the front, the air is warmer and more buoyant. During the winter season, the polar front is generally located at lower latitudes and is more pronounced than in the summer.

Cold fronts occur at the leading edge of equatorward-moving polar air. In contrast, warm fronts are well defined at the equatorward surface position of polar air as it retreats on the eastern sides of extratropical cyclones. Equatorward-moving air behind a cold front occurs in pools of dense high pressure known as polar highs and arctic highs. The term arctic high is used to define air that originates even deeper within the high latitudes than polar highs.

When polar air neither retreats nor advances, the polar front is called a stationary front. In the occluded stage of the life cycle of an extratropical cyclone, when cold air west of the surface low-pressure centre advances more rapidly toward the east than cold air ahead of the warm front, warmer, less-dense air is forced aloft. This frontal intersection is called an occluded front. Without exception, fronts of all types follow the movement of colder air.

Clouds and often precipitation occur on the poleward sides of both warm and stationary fronts and whenever tropical air reaching the latitude of the polar front is forced upward over the colder air near the surface. Such fronts are defined as active fronts. Rain and snowfall from active fronts form a major part of the precipitation received in the middle and high latitudes. Precipitation in these areas occurs primarily during the winter months.

The position of the polar front slopes upward toward colder air. This occurs because cold air tends to undercut the warmer air of tropical origin. Since cold air is more dense, atmospheric pressure decreases more rapidly with height on the poleward side of the polar front than on the warmer tropical side. This creates a large horizontal temperature contrast, which is essentially a large pressure gradient, between the polar and tropical air. In the middle and upper parts of the troposphere, this pressure gradient is responsible for the strong westerly winds occurring there. Winds created aloft circulate around a large region of upper-level low pressure near each of the poles. The centre of each low pressure region is a persistent cyclone known as the circumpolar vortex.

The region of strongest winds, which occurs at the juncture of the tropical and polar air masses, is called the jet stream. Since the temperature contrast between the tropics and the high latitudes is greatest in the winter, the jet stream is stronger during that season. In addition, since the mid-latitudes also become colder during the winter, while tropical temperatures remain relatively unchanged, the westerly jet stream approaches latitudes of 30° during the colder season. During the warmer season in both hemispheres, the jet stream moves poleward and is located between latitudes of 50° and 60°.

The jet stream reaches its greatest velocity at the tropopause. Above that level, a reversal of the horizontal temperature gradient occurs, which produces a reduction in the wind speeds of the jet stream at high latitudes. This causes a weakening of the westerlies with increasing height. At intervals ranging from 20 to 40 months, with a mean value of 26 months, westerly winds in the stratosphere reverse direction over low latitudes, so that an easterly flow develops. This feature is called the quasi-biennial oscillation (QBO). In addition, a phenomenon called sudden stratospheric warming, apparently the result of strong downward air motion, also occurs in the late winter and spring at high latitudes. Sudden stratospheric warming can significantly alter temperature-dependent chemical reactions of ozone and other reactive gases in the stratosphere and affect the development of such features as “ozone holes.”

A major focus of weather forecasting in the middle and high latitudes is to forecast the movement and development of extratropical cyclones, polar and arctic highs, and the location and intensity of subtropical ridges. Spring and fall frosts, for example, are associated with the equatorward movement of polar highs behind a cold front, while droughts and heat waves in the summer are associated with unusually strong subtropical ridges.

Effect of Continents on Air Movement

Preferred geographic locations exist for subtropical ridges and for the development, movement, and decay of extratropical cyclones. During the winter months in middle and high latitudes, the lower parts of the troposphere over continents often serve as reservoirs of cold air as heat is radiated into space throughout the long nights. In contrast, the oceans lose heat less rapidly, because of the large heat capacity of water, their ability to overturn as the surfaces cool and become negatively buoyant, and the movement of ocean currents such as the Gulf Stream and the Kuroshio current. Warm currents transport heat from lower latitudes poleward and tend to occur on the western sides of oceans. The lower troposphere over these warmer oceanic areas tends to be a region of relative low pressure. As a result of this juxtaposition of cold air and warm air, the eastern sides of continents and the western fringes of oceans in middle and high latitudes are the preferred locations for extratropical storm development. Over Asia in particular, the cold high-pressure system is sufficiently permanent that a persistent offshore flow called the winter monsoon occurs.

An inverse type of flow develops in the summer as the continents heat more rapidly than their adjacent oceanic areas. Continental areas tend to become regions of relative low pressure, while high pressure in the lower troposphere becomes more prevalent offshore. As the winds travel from areas of higher pressure to areas of lower pressure, a persistent onshore flow develops over large landmasses in the lower troposphere. The result of this heating is referred to as the summer monsoon. The leading edge of this monsoon is associated with a feature called the monsoon trough, a region of low atmospheric pressure at sea level. Tropical moisture carried onshore by the summer monsoon often results in copious rainfall. The village of Cherrapunji in northeastern India, for instance, recorded over 9 metres (about 30 feet) of rain in one month (July 1861) owing to the Indian summer monsoon.

As a result of the continental effect, the subtropical ridge is segmented into surface high-pressure cells. In the summer, large landmasses in the subtropics tend to be centres of relative low pressure as a result of strong solar heating. As a consequence, persistent high-pressure cells, such as the Bermuda and Azores highs, occur over the oceans. The oval shape of these high-pressure cells creates a thermal structure on their eastern sides that differs from the thermal structure on their western sides in the lower troposphere. On the eastern side, subsidence from the Hadley circulation is enhanced by the tendency of air to preserve its angular momentum on the rotating Earth. Owing to the enhanced descent of air over the eastern parts of the oceans, landmasses adjacent to these areas (typically the western sides of continents) tend to be deserts, such as those found in northwestern and southwestern Africa and along western coastal Mexico.

Effect of Oceans on Air Movement

The arid conditions found along the western coasts of continents in subtropical latitudes are further enhanced by the influence of the equatorward surface air flow on the ocean currents. This flow exerts a shearing stress on the ocean surface, which results in the deflection of the upper layer of water above the thermocline to the right in the Northern Hemisphere and to the left in the Southern Hemisphere. (This deflection is also the result of the Coriolis effect; water from both hemispheres moves westward when displaced toward the Equator.) As warmer surface waters are carried away by this offshore ocean airflow, cold water from below the thermocline rises to the surface in a process called upwelling. Upwelling creates areas of cold coastal surface waters that stabilize the lower troposphere and reduce the chances for convection. Lower convection in turn reduces the likelihood for precipitation, although fogs and low stratus clouds are common. Upwelling regions are also associated with enriched sea life, as oxygen and organic nutrients are transported upward from the depths toward the surface of the ocean.

During periods when the intertropical convergence zone (ITCZ) is located near the Equator, trade winds from the northeast and southeast converge there. The westward-moving winds cause the displacement of surface ocean waters away from the Equator such that the deeper, colder waters move to the surface. In the central and eastern Pacific Ocean near the Equator, when this upwelling is stronger than average, the event is called La Niña. When the trade winds weaken in this region, however, warmer-than-average surface conditions occur, and upwelling is weaker than usual. This event is called El Niño. Changes in ocean surface temperatures caused by El Niño significantly affect where cumulonimbus clouds form in the ITCZ and, therefore, the geographic structure of the Hadley cell. During periods when El Niño is active, weather patterns across the entire Earth are substantially altered.

Mountain Barriers

North-south-oriented mountain barriers, such as the Rockies and the Andes, and large massifs, such as the Plateau of Tibet, also influence atmospheric flow. When the general westerly flow in the mid-latitudes reaches these barriers, air tends to be blocked. It is transported poleward west of the terrain and toward the Equator east of the obstacle. Air forced up the slopes of mountain barriers is often sufficiently moist to produce considerable precipitation on windward sides of mountains, whereas subsiding air on the lee slopes produces more-arid conditions. Essentially, the elevated terrain affects the atmosphere as if it were an anticyclone, a centre of high pressure. In addition, mountains prevent cold air from the continental interior from moving westward of the terrain. As a result, relatively mild weather occurs along the western coasts of continents with north-south mountain ranges when compared with continental interiors. For example, the West Coast of North America experiences milder winter weather than the Great Plains and Midwest, both of which occur at similar latitudes. In contrast, east-west mountain barriers, such as the Alps in Europe, offer little impediment to the general westerly flow of air. In these situations, milder maritime conditions extend much farther inland.

Biosphere

The biosphere – the sphere of life – was named by Eduard Suess in 1875 but not fully described as a concept until the work of Vladimir Vernadsky in the 1920s. The biosphere is made up of biomes, or biophysical zones, filled with many ecosystems. Each ecosystem is composed of an intricate set of species adapted to prevailing conditions, from below ocean floors, to the land surface, to above the highest mountains. It includes life forms ranging in size from microscopic bacteria to the gargantuan blue whale. While persistent for billions of years, the biosphere has been hit by five mass extinctions in the geologic past and now faces existential threats to species diversity from human activity.

Throughout the evolution of life on Earth, from primitive organisms to the present set, all life forms have found ways to obtain energy, acquire nutrients to build organic molecules, and reproduce. Energy from the sun is captured by photosynthesizing organisms called autotrophs, or producers, that can harness solar energy to convert inorganic molecules into organic molecules -- the building blocks of life. These organic molecules store energy and are consumed by other non-photosynthetic organisms called heterotrophs, or consumers. This seemingly simple process -- grass being eaten by deer, for example -- took billions of years to develop. Through the process of evolution, species diversify to fill the available opportunities for existence, creating an ever changing set of plants and animals found in the Earth's biomes from tundra to rainforests.

Diversity of Life

Life is ubiquitous on Earth, yet biological productivity varies greatly from deserts to rainforests. Some 1.9 million species have been named, but there are an estimated five to thirty million or more species making up the biodiversity of Earth. Many of the unidentified species are in particularly hard to get to places, such as Antarctic ocean environments, or extremophiles living where it is intensely hot or cold or acidic. Below foot and beneath the sea, thousands if not millions of different organisms are teeming, many are unidentified as there just hasn't been enough time and attention to sort out all these small life forms.

Phytomass, or the mass of plants, is estimated to be about 500 to 800 GtC (billion tons of carbon). Estimates of the mass of heterotrophs are dominated by large uncertainties regarding the mass of organisms living in the soil, deep below the soil, and in ocean sediments. Prokaryotes, simple organisms without a nucleus (bacteria and archaea), alone may be equal in mass to that of plants. Land and ocean heterotrophs other than prokaryotes make up a relatively small contribution to the total mass. Estimates of the total mass of the biosphere are more than 1 TtC (trillion tons of C) and perhaps as much as 4 TtC.

The suite of species on Earth at any given time is continually changing through the process of evolution. Over geologic time, more species have gone extinct than exist today. A dramatic example of this change is past extinction events. Paleobiologists and geologists have pieced together evidence in the geologic record of five mass extinction events reducing the Earth's biodiversity to a portion of its full potential. A notable example is the mass extinction 65 million years ago that coincided with the end of the age of dinosaurs.

Abrupt change in the physical and chemical factors fundamental to life are key in mass extinctions. After each mass extinction the diversity of life slowly recovers to fill the ecospace available in the Earth's environment. This process can take millions of years of evolution.

Potential causes for mass extinctions in the past include massive and sustained volcanic eruptions and impacts from comets and/or asteroids – both causing consequent alteration of the atmosphere from the lofted debris that blocks incoming sunlight. Sustained or very rapid climate change and sea-level change are also possible explanations of past mass extinctions.

Since the last ice age, human activity changing land use has been a dramatic factor in the disruption of species habitat. For example, about 35% of ice-free land is devoted to human agriculture, and as a consequence of this expansion, species are forced into environments in which they are ill suited to survive. Since the industrial revolution, human activity has altered air and water quality and is forcing a change in climate. These factors, in addition to land use change, interact in complex ways to affect biodiversity. Many scientists consider our present age a 6th mass extinction. An estimate of extinction from future climate change (projected in the range of 3.6 to 5.4 deg F) when compounded by other human impacts to biodiversity, finds that 20 to 30 percent of the identified species known today could be lost along with the ecosystem services they provide.

Cryosphere

The cryosphere is the sum of frozen water around the globe. By volume, the perennial ice-containing cold regions of the world are dominated by the continental ice sheets of Greenland and Antarctica. Sea ice and snow have much less volume but are large in areal extent. Ice sheets, snow, and sea ice play a critical role in the Earth's climate. The sunlight reflectivity, or albedo, of these vast areas of the Earth's surface are a key component of the Earth's energy balance. Large swings in the volume of water trapped as ice occur during glacial-interglacial cycles. During periods of warming, large volumes of fresh water flow into the ocean and affect ocean salinity and currents. Ice age cycles radically alter sea level by roughly 100 meters, dramatically changing coastlines. Permafrost covers large high latitude areas and hosts its own unique ecosystems. Many forms of life from polar bears to humans of the circumpolar region have adapted to life in and near the cryosphere. As the permafrost melts with higher temperatures, scientists are concerned that organic material now locked in ice will decompose releasing carbon dioxide and methane exacerbating global warming.

The state of the cryosphere – from change in mountain glaciers to the annual extent of sea ice – serves as a readily observable marker of change in climate, both historically and currently. Today, most mountain glaciers around the world are retreating. Summer sea ice extent in the Arctic is diminishing as is the thickness of old sea ice. Melting sea ice does not affect sea level rise directly but is critical as part of the Earth's energy balance. Loss of sea ice exposes the dark ocean surface, enabling it to absorb solar radiation and warm more than it would if covered in ice. However, the melting of the great ice sheets and mountain glaciers on land do contribute to sea level rise as the melt waters make their way into the ocean. During the last glaciation, sea level is estimated to be about 120m lower than today, enlarging significantly the area of continents and allowing for organisms to cross land bridges now underwater.

Geosphere

The geosphere is the earth itself: the rocks, minerals, and landforms of the surface and interior. Below the crust - which varies in depth from about 5 km beneath the ocean floor to up to 70 km below the land surface, temperatures are high enough for deformation and a paste-like flow of elements. At one time, roughly 200 million years ago, the continents were joined together in a supercontinent called Pangaea, but since then tectonic plates have slowly separated, creating the arrangement of the continents we are accustomed to today.

Plate tectonic movement is ongoing, and humans can witness its sometimes violent activity in the form of earthquakes and volcanoes. More regularly, however, human interaction with the dynamic geosphere comes in the form of surface erosion, our use of arable land for farming, and excavations for the construction of buildings, roads, and mines.

Hydrosphere

Hydrosphere is the discontinuous layer of water at or near Earth's surface. It includes all liquid and frozen surface waters, groundwater held in soil and rock, and atmospheric water vapour.

Water is the most abundant substance at the surface of Earth. About 1.4 billion cubic km (326 million cubic miles) of water in liquid and frozen form make up the oceans, lakes, streams, glaciers, and groundwaters found there. It is this enormous volume of water, in its various manifestations, that forms the discontinuous layer, enclosing much of the terrestrial surface, known as the hydrosphere.

Central to any discussion of the hydrosphere is the concept of the water cycle (or hydrologic cycle). This cycle consists of a group of reservoirs containing water, the processes by which water is transferred from one reservoir to another (or transformed from one state to another), and the rates of transfer associated with such processes. These transfer paths penetrate the entire hydrosphere, extending upward to about 15 km (9 miles) in Earth's atmosphere and downward to depths on the order of 5 km (3 miles) in its crust.

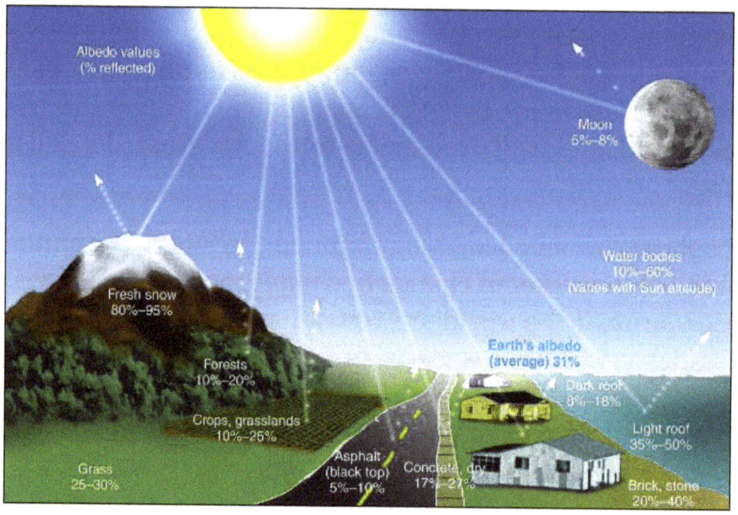

Distribution and Quantity of Earth's Waters

Ocean waters and waters trapped in the pore spaces of sediments make up most of the present-day hydrosphere. The total mass of water in the oceans equals about 50 percent of the mass of sedimentary rocks now in existence and about 5 percent of the mass of Earth's crust as a whole. Deep and shallow groundwaters constitute a small percentage of the total water locked in the pores of sedimentary rocks—on the order of 3 to 15 percent. The amount of water in the atmosphere at any one time is trivial, equivalent to roughly 13,000 cubic km (about 3,100 cubic miles) of liquid water, or about 0.001 percent of the total at Earth's surface. This water, however, plays an important role in the water cycle.

At present, ice locks up a little more than 2 percent of Earth's water and may have accounted for as much as 3 percent or more during the height of the glaciations of the Pleistocene Epoch (2.6 million to 11,700 years ago). Although water storage in rivers, lakes, and the atmosphere is small, the rate of water circulation through the rain-river-ocean-atmosphere system is relatively rapid. The amount of water discharged each year into the oceans from the land is approximately equal to the total mass of water stored at any instant in rivers and lakes.

Soil moisture accounts for only 0.005 percent of the water at Earth's surface. It is this small amount of water, however, that exerts the most direct influence on evaporation from soils. The biosphere, though primarily H_2O in composition, contains very little of the total water at the terrestrial surface, only about 0.00004 percent, yet the biosphere plays a major role in the transport of water vapour back into the atmosphere by the process of transpiration.

Earth's waters are not pure H_2O but contain dissolved and particulate materials. Thus, the masses of water at Earth's surface are major receptacles of inorganic and organic substances, and water movement plays a dominant role in the transportation of these substances about the planet's surface.

Biogeochemical Properties of the Hydrosphere

Rainwater

About 107,000 cubic km (nearly 25,800 cubic miles) of rain fall on land each year. The total water in the atmosphere is 13,000 cubic km, and this water, owing to precipitation and evaporation, turns over every 9.6 days. Rainwater is not pure but rather contains dissolved gases and salts, fine-ground particulate material, organic substances, and even bacteria. The sources of the materials in rainwater are the oceans, soils, fertilizers, air pollution, and fossil fuel combustion.

It has been observed that rains over oceanic islands and near coasts have ratios of major dissolved constituents very close to those found in seawater. The discovery of the high salt content of rain near coastlines was somewhat surprising because sea salts are not volatile, and it might be expected that the process of evaporation of water from the sea surface would "filter" out the salts. It has been demonstrated, however, that a large percentage of the salts in rain is derived from the bursting of small bubbles at the sea surface due to the impact of rain droplets or the breaking of waves, which results in the injection of sea aerosol into the atmosphere. This sea aerosol evaporates, with resultant precipitation of the salts as tiny particles that are subsequently carried high into the atmosphere by turbulent winds. These particles may then be transported over continents to fall in rain or as dry deposition.

Assuming equilibrium with the atmospheric carbon dioxide partial pressure (PCO_2) of 10−3.5 (0.00035) atmosphere, the approximate mean composition of rainwater is in parts per million (ppm): sodium (Na^+), 1.98; potassium (K^+), 0.30; magnesium (Mg_2^+), 0.27; calcium (Ca_2^+), 0.09; chloride (Cl^-), 3.79; sulfate (SO_4^{2-}), 0.58; and bicarbonate (HCO_3^-), 0.12. In addition to these ions, rainwater contains small amounts of dissolved silica—about 0.30 ppm. The average pH value of rainwater is 5.6. (The term pH is defined as the negative logarithm of the hydrogen ion concentration in moles per litre. The pH scale ranges from 0 to 14, with lower numbers indicating increased acidity.) On a global basis, as much as 35 percent of the sodium, 55 percent of the chlorine, 15 percent of the potassium, and 37 percent of the sulfate in river water may be derived from the oceans through sea aerosol generation.

A considerable amount of data has become available for marine aerosols. These aerosols are important because: (1) they are vital to any description of the global biogeochemical cycle of an element, (2) they may have an impact on climate, (3) they are a sink, via heterogeneous chemical reactions, for trace atmospheric gases, and (4) they influence precipitation of cloud and rain droplets. For many trace metals, the ratio of the atmospheric flux to the riverine flux for coastal and remote oceanic areas may be greater than one, indicating the importance of atmospheric transport. Figures have been prepared that illustrate the enrichment factors (EF) of North Atlantic marine aerosols and suspended matter in North Atlantic waters relative to the crust (that is, terrestrial sources), where:

$$\mathrm{EF}_{crust} = \frac{(X/\mathrm{Al})_{air}}{(X/\mathrm{Al})_{crust}},$$

and (X/Al)air and (X/Al)crust refer, respectively, to the ratio of the concentration of the element X to that of Al, aluminum (which is an easily observed terrestrial component of aerosols), in the atmosphere and in average crustal material. Comparing the enrichment factors in marine aerosols with those of suspended matter in the water column indicates qualitatively the marine aerosols' importance as a source that alters the composition of marine suspended matter and, consequently, their importance to deep-sea sedimentation. Moreover, such comparisons help identify how significant terrestrial sources are for both the marine aerosols and the water below.

In some instances the ratios of ions in rainwater deviate significantly from those in seawater. Mechanisms proposed for this fractionation are, for example, the escape of chlorine as gaseous hydrogen chloride (HCl) from sea salt aerosol with a consequent enrichment in sodium and bubbling and thermal diffusion. In addition, release of biogenic gases such as dimethyl sulfide (DMS) from the sea surface and its subsequent reaction in the oceanic atmosphere to sulfate can change rainwater ion ratios with respect to seawater. Soil particles also can influence rainwater composition. Rainfall over the southwestern United States contains relatively high sulfate concentrations because of sulfate-bearing particles that have been blown into the atmosphere from desert soils. Rain near industrial areas commonly contains high contents of sulfate, nitrate, and carbon dioxide (CO_2) largely derived from the burning of coal and oil. There are two main processes leading to the conversion of sulfur dioxide (SO_2) to sulfuric acid (H_2SO_4). These are reactions with hydroxyl radicals (–OH) and with hydrogen peroxide (H_2O_2) in the atmosphere:

$$\mathrm{SO_2 + OH \rightarrow intermediate\ species \rightarrow H_2SO_4}$$

and

$$\mathrm{SO_2 + H_2SO_2 = H_2SO_4}$$

The sulfuric acid then dissociates to hydrogen and sulfate ions:

$$H_2SO_4 = 2H^+ + SO_4^{2-}$$

For the nitrogen gases nitric oxide (NO) and nitrogen dioxide (NO_2) released from fossil fuel burning, their atmospheric reactions lead to the production of nitric acid (HNO_3) and its dissociation to hydrogen ions (H^+) and nitrate (NO_3^-). These reactions are responsible for the acid rain conditions that occurred in the northeastern United States, southeastern Canada, and western Europe during the second half of the 20th century. The high sulfate values of the rain in the northeastern United States reflect the acid precipitation conditions of this region.

River and Ocean Waters

River discharge constitutes the main source for the oceans. Seawater has a more uniform composition than river water. It contains, by weight, about 3.5 percent dissolved salts, whereas river water has only 0.012 percent. The average density of the world's oceans is roughly 2.75 percent greater than that of typical river water. Of the average 35 parts per thousand salts of seawater, sodium and chlorine make up almost 30 parts, and magnesium and sulfate contribute another four parts. Of the remaining one part of the salinity, calcium and potassium constitute 0.4 part each and carbon, as carbonate and bicarbonate, about 0.15 part. Thus, nine elements (hydrogen, oxygen, sulfur, chlorine, sodium, magnesium, calcium, potassium, and carbon) make up 99 percent of seawater, though most of the 94 naturally occurring elements have been detected therein. Of importance are the nutrient elements phosphorus, nitrogen, and silicon, along with such essential micronutrient trace elements as iron, cobalt, and copper. These elements strongly regulate the organic production of the world's oceans.

A portion of the delta of the Mekong River as it flows through southern Vietnam and empties into the South China Sea.

In contrast to ocean water, the average salinity of the world's rivers is low—only about 0.012 percent, or 120 ppm by weight. Of this salt content, carbon as bicarbonate constitutes 58 parts, or 48 percent, and calcium, sulfur as sulfate, and silicon as dissolved monomeric silicic acid make up a total of about 39 parts, or 33 percent. The remaining 19 percent consists predominantly of chlorine, sodium, and magnesium in descending importance. It is obvious that the concentrations and relative proportions of dissolved species in river waters contrast sharply with those of seawater. Thus, even though seawater is derived in part by the chemical differentiation and evaporation of river water, the processes involved affect every element differently, indicating that simple evaporation and concentration are entirely secondary to other processes.

Water-rock Interactions as Determining River Water Composition

Generally speaking, the composition of river water, and thus that of lakes, is controlled by water-rock interactions. The attack of carbon dioxide-charged rain and soil waters on the individual minerals in continental rocks leads to the production of dissolved constituents for lakes, rivers, and streams. It also gives rise to solid alteration products that make up soils or suspended particles in freshwater aquatic systems. The carbon dioxide content of rain and soil waters is of particular importance in weathering processes. The pH of rainwater equilibrated with the atmospheric carbon dioxide partial pressure of 10−3.5 (0.00032) atmosphere is 5.6. In industrial regions, rainwater pH values may be lower because of the release and subsequent hydrolysis of acid gases—namely, sulfur dioxide and nitrogen oxides (NO_x) from the combustion of fossil fuels. After rainwater enters soils, its characteristics change markedly. The usual few parts per million of salts in rainwater increase substantially as the water reacts. The upper part of the soil is a zone of intense biochemical activity. The bacterial population near the surface is large, but it decreases rapidly downward. One of the major biochemical processes of the bacteria is the oxidation of organic material, which leads to the release of carbon dioxide. Soil gases obtained above the zone of water saturation may contain 10 to 40 times as much carbon dioxide as the free atmosphere, and in some cases carbon dioxide has been shown to make up 30 percent of the soil gases as opposed to 0.03 percent of the free atmosphere. In addition to the acid effects of carbon dioxide, there is a highly acidic micro-environment created by the roots of living plants. Values of pH as low as 2 have been measured immediately adjacent to root hairs. The combined length of a plant's root hairs may be several kilometres, so their chemical effects on acidic water are formidable.

Suspended soil particles. The Amazon River near Manaus, Brazil. The brown colour of the water is the result of rains washing soil particles into the stream from surrounding land and the stirring up of mud on the riverbed.

Congruent and Incongruent Weathering Reactions

These acid solutions in the soil environment attack the rock minerals, the bases of the system, producing neutralization products of dissolved constituents and solid particles. Two general types of reactions occur: congruent and incongruent. In the former, a solid dissolves, adding elements to the water according to their proportions in the mineral. An example of such a weathering reaction is the solution of calcite ($CaCO_3$) in limestones:

$$CaCO_3 + CO_2(g) + H_2O = Ca^{2+} + 2HCO_3^-$$

Here one of the HCO_3^- ions comes from calcite and the other from $CO_2(g)$ in the reacting water. The amount of carbon dioxide dissolved according to reaction $CaCO_3 + CO_2(g) + H_2O = Ca^{2+} + 2HCO_3^-$ depends on temperature, pressure, original bicarbonate content of the weathering solution, and partial pressure of the carbon dioxide. The carbon dioxide and the temperature are the most important variables. Increases in one or both of these variables lead to increases in the amount of calcite dissolved. For example, for a carbon dioxide pressure of 10−3.5 (0.00032) atmosphere, the amount of calcium that can be dissolved until saturation is about 10−3.3 (0.0005) mole, or 20 ppm, at 25 °C (77 °F). For an atmospheric carbon dioxide pressure of 10−2 (0.01) atmosphere and for a soil atmosphere of nearly pure carbon dioxide, the values are 65 and 300 ppm, respectively. The weathering of calcite leads to the release of calcium and bicarbonate ions into soil waters and groundwaters, and these constituents eventually reach lake and river systems. The insoluble residue of quartz (SiO_2), clay minerals, and iron oxides (e.g., FeOOH) in the limestone rock make up the deep-red soils that form from limestone weathering. These particles may be carried into streams by runoff and hence to lakes and the oceans and become part of the suspended load of these systems.

An example of an incongruent weathering reaction—that is, one where only part of a solid is consumed—is that involving aluminosilicates. One such reaction is the aggressive attack of carbon dioxide-charged soil water on the mineral K-spar ($KAlSi_3O_8$), an important phase found in continental rocks. The reaction is:

$$2KA1Si_3O_8 + 2CO_2 + 11H_2O = \\ A1_2Si_2O_5(OH)_4 + 2K^+ + 2HCO_3^- + 4H_4SiO_4$$

It should be noted that the K-spar changes into a new mineral—kaolinite (a clay mineral) in this case—plus solution, and acid is consumed. The total dissolved material per litre of soil solution released is about 60 ppm for a solution initially equilibrated with a typical soil carbon dioxide content. The water resulting from reaction above would contain bicarbonate, potassium, and dissolved silica in the ratios 1:1:2, and the new solid, kaolinite, would be a weathering product. These dissolved constituents and the solid alteration product would eventually reach rivers to be transferred possibly to lakes and ultimately to the sea. It has been demonstrated that the composition of river water is the product of a variety of mineral-water reactions. The dissolved load of the world's rivers comes from the following sources: 7 percent from beds of halite (NaCl) and salt disseminated in rocks, 10 percent from gypsum ($CaSO_4 \cdot 2H_2O$) and anhydrite ($CaSO_4$) deposits and sulfate salts disseminated in rocks, 38 percent from limestones and dolomites, and 45 percent from the weathering of one silicate mineral to another. Of the bicarbonate ions in river water, 56 percent stems from the atmosphere, 35 percent from carbonate minerals, and 9 percent from the oxidative weathering of fossil organic matter. Reactions involving silicate minerals account for 30 percent of the riverine bicarbonate ions.

Besides dissolved substances, rivers also transport solids in traction (i.e., bed load) and, most importantly, suspended load. The present global river-borne flux of solids to the oceans is estimated as 15.5 billion metric tons (about 17.1 billion tons) per year. Present elemental fluxes are estimated in millions of metric tons per year as silicon, 4,420; aluminum, 1,460; iron, 740; calcium, 330; potassium, 310; magnesium, 210; and sodium, 110. The total load of particulate organic carbon of the world's rivers is 180 million metric tons (nearly 200 million tons) per year. The riverine fluxes of trace metals to the oceans are dominated by their occurrence in the particulate phase as opposed to the dissolved phase. The particulate matter in river water is an important source of silicon, aluminum, iron, titanium, rubidium, scandium, vanadium, the lanthanoids, and other elements for deep-sea sediments.

Lake Waters

Although lake waters constitute only a small percentage of the water in the hydrosphere, they are an important ephemeral storage reservoir for fresh water. Aside from their recreational use, lakes constitute a source of water for household, agricultural, and industrial uses. Lake waters are also very susceptible to changes in chemical composition due to these uses and to other factors.

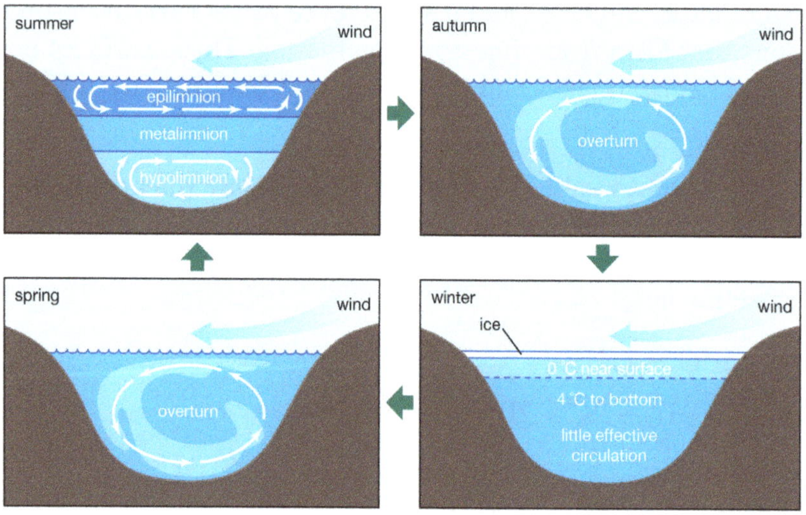

Temperate Lake Circulation

Annual circulation patterns in a dimictic lake. The typical dimictic lake has distinct layers that fully mix twice a year. It undergoes stratification in the summer and complete overturn in the autumn and spring. During winter, surface ice prevents further mixing by the wind. Small differences in density and temperature exist, with cooler water (0 °C [32 °F]) staying near the surface and warmer, denser water (4 °C [39.2 °F]) extending to the bottom.

In general, fresh waters at the continental surface evolve from their rock sources by enrichment in calcium and sodium and by depletion in magnesium and potassium. In very soft waters the alkalies may be more abundant than the alkaline earths, and in the more-concentrated waters of open river systems calcium > magnesium > sodium > potassium. For the anions, in general, HCO_3^- exceeds SO_4^{2-}, which is greater in concentration than Cl^-. It is worthwhile at this stage to consider some major mechanisms that control global surface water composition. These mechanisms are atmospheric precipitation, rock reactions, and evaporation-precipitation.

The mechanism principally responsible for waters of very low salinity is precipitation. These waters tend to form in tropical regions of low relief and thoroughly leached source rocks. In these regions rainfall is high, and volumes of fresh water (rivers, tributary streams, pools, etc.) within a watershed are usually dominated by salts brought in by precipitation. Such waters constitute one part of a chain of water volumes that begins with falling precipitation and ends with the release of water into the ocean, for which the final part of the chain represents water volumes dominated by contributions of dissolved salts from the rocks and soils of their basins. These waters have moderate salinity and are rich in dissolved calcium and bicarbonate. They are in turn the "end-member" of another series that extends from the calcium-rich, medium-salinity fresh waters to the high-salinity, sodium chloride-dominated waters of which seawater is an example. Seawater composition, however, does not evolve directly from the composition of fresh waters and the precipitation of calcium carbonate; other mechanisms that control its composition are involved. Such factors as relief and vegetation also may affect the composition of the world's surface waters, but atmospheric precipitation, water-rock reactions, and evaporation-crystallization processes appear to be the dominant mechanisms governing continental surface water chemistry.

Continental fresh waters evaporate once they have entered closed basins, and their constituent salts precipitate on the basin floors. The composition of these waters may evolve along several different paths, depending on their initial chemical makeup.

Biological processes strongly affect the composition of lake waters and are responsible to a significant degree for the compositional differences between the upper water layer (the epilimnion) and the lower water layer (the hypolimnion) of lakes. The starting point is photosynthesis, represented by the following reaction:

$$106CO_2 + 16NO_3^- + HPO_4^{2-} + 122H_2O + 18H^+ \xrightarrow[\text{micro-nutrients}]{\text{radiant energy}}$$

$$C_{106}H_{263}O_{110}N_{16}P + 138O_2$$

The reversal of this reaction is oxidation-respiration leading to the release of the nutrients nitrogen and phosphorus, as well as carbon dioxide. In a stratified lake, carbon, nutrients, and silica are extracted from the upper layer during photosynthesis. This process leads to reduced concentrations of nitrate, phosphate, and silica in these waters and, during times of maximum daylight organic production, to supersaturation of the upper layer with respect to dissolved oxygen. The organic matter produced by phytoplankton may be either grazed upon by zooplankton and other organisms or decomposed by bacteria. Some of it, however, sinks into the lower layer. There it is further decomposed, especially by bacteria, resulting in the release of dissolved phosphorus and nitrogen and the consumption of oxygen. Oxygen concentrations therefore are reduced in these lower lake waters, because stratification prevents oxygen exchange with the atmosphere. Furthermore, the inorganic carbonate and siliceous skeletons of the dead organisms sinking into the lower layer may dissolve, giving rise to increased concentrations of dissolved silica and inorganic carbon in the deep waters of stratified lakes. This dissolution is a result of undersaturation of the waters of the lower layer with respect to the opaline silica and calcium carbonate that make up the skeletons of the dead and sinking plankton. These natural biological processes have been accelerated in some lakes because of excess nutrient input by human activity, resulting in the eutrophication of lake waters and marine systems.

Groundwaters

Groundwaters derive their compositions from a variety of processes, including dissolution, hydrolysis, and precipitation reactions; adsorption and ion exchange; oxidation and reduction; gas exchange between groundwater and the atmosphere; and biological processes.

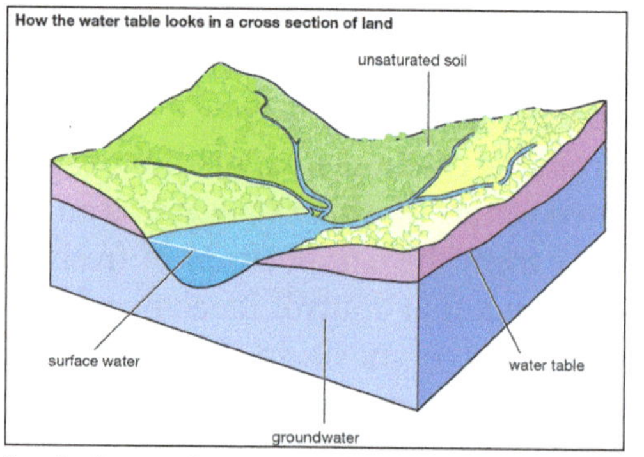

The water table is the top level of groundwater. Surface water is an exposed part of the water table.

The biological processes of greatest importance are microbial metabolism, organic production, and respiration (oxidation). By far the most important overall process for the major constituents of groundwater is that of mineral-water reactions, which were briefly described above in River and ocean waters. Thus, the composition of groundwaters strongly reflects the types of rock minerals that the waters have encountered in their movement through the subsurface.

In general, the most mobile elements in groundwater—i.e., those most easily liberated by the weathering of rock minerals—are calcium, sodium, and magnesium. Silicon and potassium have intermediate mobilities, and aluminum and iron are essentially immobile and locked up in solid phases.

Groundwaters are highly susceptible to contamination because of human activities and the fact that their dissolved constituents are derived to a large extent from the leaching of surface materials. Some of the nitrogen and phosphorus applied to soils as fertilizers and organic pesticides may be leached and leak into groundwater systems, leading to increased concentrations of ammonium and phosphate. Radioactive wastes, industrial chemicals, household materials, and mine refuse are other anthropogenic sources of dissolved substances that have been detected in groundwater systems.

Ice

Ice is nearly a pure solid and, as such, accommodates few foreign ions in its structure. It does contain, however, particulate matter and gases, which are trapped in bubbles within the ice. The change in composition of these materials through time, as recorded in the successive layers of ice, has been used to interpret the history of Earth's surface environment and the impact of human activities on this environment. The increase in the lead content of continental glacial ice with decreasing age of the ice up to the middle of the 1970s, for example, reflects the progressive input of tetraethyl lead into the global environment from gasoline burning. (Stringent environmental regulations that appeared in the 1970s regarding the use of leaded gasoline has led to a fall in lead concentrations in ice laid down since that time.) Also, atmospheric carbon dioxide and methane concentrations, which have increased significantly during the past century because of anthropogenic activities, are faithfully recorded in ice bubbles of the thick continental ice sheets. By 2016 atmospheric carbon dioxide and methane concentrations had increased by more than 43 percent and more than 150 percent, respectively, higher than their concentrations 200 years ago; the latter concentration values were obtained from measurements of the gases in air trapped in ice.

Perito Moreno Glacier, Los Glaciares National Park

The Water Cycle

General Nature of the Cycle

The present-day water cycle at Earth's surface is made up of several parts. Some 496,000 cubic km (about 119,000 cubic miles) of water evaporates from the land and ocean surface annually, remaining for about 10 days in the atmosphere before falling as rain or snow. The amount of solar radiation necessary to evaporate this water is half of the total solar radiation received at Earth's surface. About one-third of the precipitation falling on land runs off to the oceans primarily in rivers, while direct groundwater discharge to the oceans accounts for only about 0.6 percent of the total discharge. A small amount of precipitation is temporarily stored in the waters of rivers and lakes. The remaining precipitation over land, 73,000 cubic km (17,500 cubic miles) per year, returns to the atmosphere by evaporation. Over the oceans, evaporation exceeds precipitation, and the net difference represents transport of watcr vapour over land, where it precipitates as rain or snow and returns to the oceans as river runoff and direct groundwater discharge.

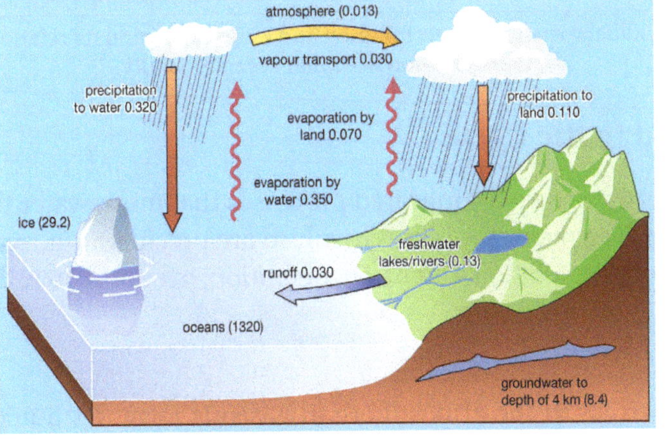

The present-day surface hydrologic cycle, in which water is transferred from the oceans through the atmosphere to the continents and back to the oceans over and beneath the land surface. The values in parentheses following the various forms of water (e.g., ice) refer to volumes in millions of cubic kilometres; those following the processes (e.g., precipitation) refer to their fluxes in millions of cubic kilometres of water per year.

The various reservoirs in the water cycle have different water residence times. Residence time is defined as the amount of water in a reservoir divided by either the rate of addition of water to the reservoir or the rate of loss from it. The oceans have a water residence time of 3,000 to 3,230 years; this long residence time reflects the large amount of water in the oceans. In the atmosphere the residence time of water vapour relative to total evaporation is only about 10 days. Lakes, rivers, ice, and groundwaters have residence times lying between these two extremes and are highly variable.

There is considerable variation in evaporation and precipitation over the globe. In order for precipitation to occur, there must be sufficient atmospheric water vapour and enough rising air to carry the vapour to an altitude where it can condense and precipitate. Precipitation and evaporation vary with latitude and their relation to the global wind belts. The trade winds, for example, are initially cool, but they warm up as they blow toward the Equator. These winds pick up moisture from the ocean, increasing ocean surface salinity and causing seawater at the surface to sink. When the trade winds reach the Equator, they rise, and the water vapour in them condenses and forms clouds. Net precipitation is high near the Equator and also in the belts of the prevailing westerlies, where there is frequent storm activity. Evaporation exceeds precipitation in the subtropics, where the air is stable, and near the poles, where the air is both stable and has a low water vapour content because of the cold. The Greenland Ice Sheet and the Antarctic Ice Sheet formed because the very low evaporation rates at the poles resulted in precipitation exceeding evaporation in these local regions.

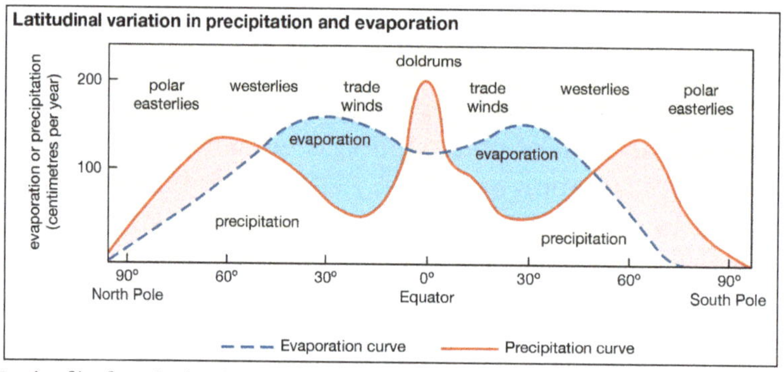

Latitudinal variation in precipitation and evaporation and its relationship to major wind belts and oceanic salinity.

Processes involved in the Cycle

The water cycle consists of various complicated processes that move water throughout the different reservoirs on the planet. The major processes involved are precipitation, evaporation, interception, transpiration, infiltration, percolation, retention, detention, overland flow, throughflow, and runoff.

Water Vapour and Precipitation

As noted above, water exists in the atmosphere in gaseous form. Its liquid form, either as water droplets in clouds or as rain, and its solid form, as ice crystals in clouds, snowflakes, or hail, occur only momentarily and locally.

Water vapour performs two major functions: (1) it is important to the radiation balance of Earth, as its presence keeps the planetary surface warmer than would otherwise be the case, and (2) it is the principal phase of the ascending part of the water cycle.

The mass of water vapour in the atmosphere, which represents only 0.001 percent of the hydrosphere, is highest in the tropics and decreases toward the poles. At a mean temperature of Earth's surface of 15 °C (59 °F), the partial pressure of water vapour at equilibrium with pure water is 0.017 atmosphere. The addition of salts to pure water lowers its vapour pressure. The equilibrium, or saturation, water vapour pressure of a saturated solution of sodium chloride is 22 percent lower than that of pure water. Precipitable water vapour has, on the average, a vapour pressure of 0.0025 atmosphere, which amounts to 15 percent of the saturation vapour pressure. The ratio of observed water vapour pressure to the saturation vapour pressure at the same temperature is the relative humidity of the air. Thus, the mean relative humidity of the atmosphere is only 15 percent, a value that is low in human terms, in which levels of 50 to 60 percent are preferred for maximum comfort. The relative humidity of the air, however, varies greatly from one geographic region to another and also vertically in the atmosphere. Atmospheric water vapour decreases rapidly with increasing altitude relative to its surface value. The amount of water required to saturate a volume of air depends on the temperature of the air. Air at high temperature can hold more water vapour at saturation than can air at low temperature. Because the temperature of the lower atmosphere (the troposphere) decreases rapidly with increasing altitude to about 15 km (about 9 miles), the upper levels of the troposphere contain little water vapour; most of the vapour is found within a few kilometres of Earth's surface. The average relative humidity of tropospheric air is about 50 percent. Above 15 km, water vapour is essentially frozen out of the atmosphere, amounting to less than 0.1 percent of its concentration at Earth's surface.

Aside from temperature, other factors determine the water vapour content of the air and are particularly important in the lower troposphere. These factors include local evaporation and the horizontal atmospheric transportation of moisture, which varies with altitude, latitude, season, and topography. During a period of 10 days (i.e., the residence time of water in the atmosphere), horizontal eddy turbulence may disperse vapour over distances up to 1,000 km (600 miles).

When a mass of air at Earth's surface is exposed to a body of water, it gains water by evaporation or loses water by precipitation, depending on its relative humidity. If the air is undersaturated, with a relative humidity of less than 100 percent, it gains water vapour because the rate of evaporation exceeds the rate of condensation. If the air is supersaturated, with a relative humidity greater than 100 percent, the air mass loses water vapour because the rate of precipitation exceeds that of evaporation. This interaction between air masses and surface water bodies drives the atmosphere toward a state of saturation, which is not achieved for the entire atmosphere because of the variability in weather and because not all air masses are in contact with water bodies. In general, the level of atmospheric water vapour is higher in the summer, since temperatures are higher at this time of year. Also, atmospheric water vapour content is higher near the source of moisture than in distant regions. Over the oceans, the air is almost always near saturation, whereas over the deserts, where the supply of moisture is limited, the air is far below water vapour saturation values. In most cases, atmospheric water vapour content decreases inland over continents, but this decrease is modified by rainfall conditions, by the presence or absence of high mountains, large lakes, extensive forests, and swamps, and by the prevailing wind directions. Horizontal winds and air mass movements transfer water vapour from the ocean to the land. Although the processes are not completely separable, the horizontal transfer of water vapour seldom causes the vapour to undergo condensation, whereas vertical movements are most important in the condensation process.

Condensation depends strongly on the average temperature of Earth's surface because the water vapour content of the air is strongly dependent on temperature. In figures that show the states of water as a function of the variables of pressure and temperature, the slope of the phase boundary between liquid water and water vapour is positive, implying that with increasing temperature the air at equilibrium will hold increasing amounts of water vapour. Cooling or mixing of this air results in condensation of the vapour and precipitation as water droplets or as ice crystals if the air temperature is below 0 °C (32 °F). When first formed, the water droplets or ice crystals are very small, on the order of 10^{-2} to 10^{-3} cm (0.004 to 0.0004 inch) in diameter, and they float freely in the atmosphere. In large quantities, these water droplets and ice crystals produce a cloud. All clouds are formed as a result of cooling below the dew point, the temperature at which condensation begins when air is cooled at constant pressure and constant water vapour content. When the droplets or crystals coalesce to a size of about 10^{-2} cm (0.004 inch) in diameter, they become heavy enough to fall as raindrops or snowflakes. Hailstones measure about 10^{-1} cm (0.04 inch) in diameter or much larger. Water vapour condensing in the atmosphere contains strongly soluble salts (mostly of oceanic origin), weakly soluble or insoluble solids (dust), and dissolved gases. The dust and sea salt aerosol particles in the air may act as sites of condensation by serving as nuclei for bringing initially a few water molecules together and inducing condensation from supersaturated air.

Distribution of Precipitation

Precipitation falling toward Earth's surface may suffer several fates. It may be evaporated during its fall or after it reaches the ground surface. If the surface is covered with dense vegetation, much of the precipitation may be held on leaves and plant limbs and stems. This process is termed interception and may result in little water reaching the ground because the water may be directly evaporated from plant surfaces back into the atmosphere. If precipitation reaches the ground in the form of snow, it may remain there for some time. On the other hand, if precipitation falls as rain, it may evaporate, infiltrate the soil, be detained in small catchment areas, or become overland flow—a form of runoff. Overland flow (R_o) may be expressed in terms of intensity units, water depth per unit of time (e.g., centimetres per hour, or inches per hour), as:

$$R_o = P - I$$

where P is precipitation rate and I is infiltration rate (rate of entry and downward movement of water into the soil profile). Infiltration rate will equal precipitation rate until the limit of the infiltration rate, or infiltration capacity, is reached. Soil infiltration rates are usually high at the beginning of a rain preceded by a dry spell and decrease as the rainfall continues. This change in rate is due to the clogging of soil pores by particles brought from above by the infiltrating rain and to the swelling of colloidal soil particles as they absorb water. Thus, rapid decreases in infiltration rates during a rain are more likely to occur in clay-rich soils than in sandy soils.

Between rainfall periods, water held in the soil as moisture is gradually lost by direct evaporation or by withdrawal by plants. Evaporation into the open atmosphere occurs at the surface of the soil, and the soil dries progressively downward with time. Water vapour in the soil diffuses upward, replenishing the evaporated water, and in turn is evaporated. The pumping of air into and out of the soil by atmospheric pressure changes enhances the movement of soil moisture upward. It has been shown that evaporation of a water droplet in the free atmosphere, and to a first approximation in various soil atmospheres, is proportional to the droplet surface area $4\pi r^2$ (square centimetres, where r is the radius of the droplet), the diffusional flux of water at the droplet surface, and the transfer of heat as the droplet evaporates. The equation for the rate of shrinkage of a water droplet due to evaporation is:

$$\frac{dr}{dt} = -\frac{D\rho_{vo}(Sp-1)}{r\rho L(1+X)},$$

where dr/dt is the rate of change in the radius of the water droplet (centimetres per second), D is the diffusion coefficient of water vapour in air (cubic centimetres per second), ρvo is the equilibrium vapour concentration at the droplet surface, Sp is the degree of undersaturation of water vapour in the environment, ρL is the density of liquid water (grams per cubic centimetre), and X is a dimensionless parameter depending on D, ρvo, temperature, the heat of evaporation of water vapour, the coefficient of thermal conductivity of air, and the spherical coordinate system necessary to define processes occurring to a spherical water droplet. Water droplets shrink—dr/dt < 0, evaporate—when the water vapour concentration in the environment (atmosphere or soil atmosphere) is less than the saturation water vapour concentration at the droplet surface. They grow—dr/dt > 0, condense—when the converse is true in the free atmosphere. The term dr/dt has negative values for evaporation and positive ones for condensation. Use of this equation shows, as an example, that it would take 23 minutes for a water droplet to shrink (evaporate) in size from 50 to 5 micrometres (0.002 to 0.0002 inch) in air at 10 °C (50 °F) and a water vapour undersaturation of 1 percent.

Besides simple evaporation of water from soils, water is also returned to the atmosphere by transpiration in plants. Plants draw water from soil moisture through their vast network of root hairs and rootlets. This water is carried upward through the plant trunk and branches into the leaves, where it is discharged as water vapour. The term evapotranspiration is used in climatic and hydrologic studies to include the combined water loss from Earth's surface resulting from evaporation and transpiration. The maximum possible evapotranspiration is termed potential evapotranspiration and is governed by the available heat energy. It is taken as equal to evaporation from a large water surface and is generally much less than actual evapotranspiration. Actual evapotranspiration is never greater than precipitation except on irrigated land because of percolation of water into groundwater bodies and surface runoff.

The soil moisture zone gains water by precipitation and infiltration and loses water by evapotranspiration, overland flow, and percolation of water downward due to gravity into the groundwater zone. The contact between the groundwater zone (phreatic zone) and the overlying unsaturated zone (vadose zone) is called the groundwater table. The water balance equation for change of moisture storage in a soil is given as:

$$S = P - E - R,$$

where S is storage, P is precipitation, E is evaporation, and R is surface runoff plus percolation rate into the groundwater zone; all terms are in units of length per unit of time (e.g., millimetres per day, centimetres per month). In humid midlatitude climates where a strong contrast between winter and summer temperatures exists, there is an annual cycle of the water content of soils. The annual cycle of moisture in soil in Ohio, U.S., for example, demonstrates the processes controlling soil moisture. Of special importance is the fact that the soils are saturated in this temperate climate in the spring, and the evaporation rate is low because of the low input of radiant energy from the

Sun. By contrast, in the summer, evaporation increases because of increasing solar radiation, and with the growth of plants so does transpiration. Soil moisture is reduced to very low levels at this time of year.

Groundwaters and River Runoff

The term R in the water balance equation for change of soil moisture storage above represents groundwater and river runoff losses from the soil moisture zone. Water percolates from the soil moisture zone through the unsaturated (vadose) zone to the water table. Flow through the unsaturated zone is complicated. After a rainfall, water may form a nearly continuous phase in pores in this zone, but, with drying, the last amount of water is held in clusters at points of contact of solid grains and as thin films on solid surfaces. The flow paths of water become more tortuous, and the water-conducting properties decrease rapidly. Structured soils and fractured rock in the vadose zone may act as conduits for fluids to reach the water table. Because of the complex geometry of water contained in the unsaturated zone, the properties of water are expressed by means of empirical relationships. Darcy's law, derived in 1856 from experimentation by the French engineer Henri Darcy, permits quantification of water flow through porous media. The law states that the rate of flow Q of a fluid through a porous layer of medium (e.g., a sand bed) is directly proportional to the area A of the layer and to the difference Δh between the fluid heads at the inlet and outlet faces of the layer and is inversely proportional to the thickness L of the layer. Expressed analytically:

$$Q = \frac{KA\Delta h}{L},$$

where K is a constant characteristic of the medium. The term K for a porous rock medium is the volume of fluid of unit viscosity passing through a unit cross section of the rock in unit time under the action of a unit pressure gradient; this characteristic is called permeability. The permeability of a rock is dependent on the geometric properties of the rock, such as porosity, shape and size distribution of constituent rock grains, and degree of cementation of the rock. Permeabilities of rocks vary greatly. Unconsolidated sands may have permeabilities measured in hundreds of darcys, whereas consolidated sands that will transmit reasonable amounts of fluid have permeabilities of 0.01 to 1 darcy. A rough idea of the meaning of one darcy of permeability (which equals 9.869×10^{-12} square metre [1.261×10^{-11} square foot]) can be obtained by imagining a cube of sand one foot on a side. If the sand has a permeability of one darcy, approximately one barrel of water per day will pass through the one-foot cube with a one-pound pressure head. The general equation of Darcy can be modified to express flow in both the unsaturated zone and the saturated groundwater zone.

Groundwater is constantly in motion. When a lake or stream intersects the groundwater table, groundwater communicates directly with these bodies of water. If the groundwater table is higher than the stream or lake level, a pressure head will develop such that the groundwater flows into the water body; conversely, if the groundwater table is lower than the river or lake level, the pressure gradient induces flow into the groundwater. Most groundwater ultimately reaches the channels of surface streams and rivers and flows to the sea. On the average, groundwater contributes to total river runoff about 30 percent of its water on a global basis.

Water runoff from the land surface is that part of precipitation which eventually appears in perennial or intermittent surface streams. Streamflow-generation mechanisms have been studied for several decades, and there is now considerable knowledge regarding rainfall runoff processes and their controls. This understanding is the result of both careful observations from field experiments and the heuristic simulations of hypothetical realities with rigorous mathematical models. The discharge measured at the downstream end of a channel reach is supplied by channel inflow at the upstream end of the reach and by the lateral inflows that enter the channel from the hillslope along the reach. The lateral inflows may arrive at the stream in one of three forms: (1) groundwater flow, (2) subsurface storm flow, or (3) overland flow.

Groundwater flow provides the base flow component of streams that sustains their flow between storms. The "flashy" response of streamflow to individual precipitation events may be ascribed to either subsurface storm flow or overland flow. Under intense rainfall events during which the surface soil layer becomes saturated to some depth, water is able to migrate through "preferred pathways" rapidly enough to deliver contributions to the stream during the peak runoff period. The conditions for subsurface storm flow are quite restrictive. The mechanism is most likely to be operative on steep, humid, forested hillslopes with very permeable surface soils.

Overland flow is generated at a point on a hillslope only after surface ponding takes place. Ponding cannot occur until the surface soil layers become saturated. It is now widely recognized that surface saturation can occur because of two quite distinct mechanisms—specifically, Horton overland flow (named for American hydraulic engineer and hydrologist Robert E. Horton) and Dunne overland flow (named for British hydrologist Thomas Dunne).

The former classic mechanism is for a precipitation rate that exceeds the saturated hydraulic conductivity of the surface soil. A moisture content versus depth profile during such a rainfall event will show moisture contents that increase at the surface as a function of time. At some point in time the surface becomes saturated, and an inverted zone of saturation begins to propagate downward into the soil. It is at this time that the infiltration rate drops below the rainfall rate and overland flow is generated. The time is called the ponding time. The necessary conditions for the generation of overland flow by the Horton mechanism are (1) a rainfall rate greater than the saturated hydraulic conductivity of the soil and (2) a rainfall duration longer than the required ponding time for a given initial moisture profile. Horton overland flow is generated from partial areas of the hillslope where surface hydraulic conductivities are lowest.

In Dunne overland flow, the precipitation rate is less than the saturated hydraulic conductivity, and the initial water table is shallow or there is a shallow impeding layer. Surface saturation occurs because of a rising water table; ponding and overland flow occur at a time when no further soil moisture storage is available. The Dunne mechanism is more common to near-channel areas. Dunne overland flow is generated from partial areas of the hillslope where water tables are shallowest. Both Horton and Dunne mechanisms result in variable source areas that expand and contract through wet and dry periods.

Total river discharge and the chemistry of the discharge vary from continent to continent; some continents are wetter and some drier than the world average, but the deviations are not extreme. The runoff per unit area from Asia and Europe is almost exactly equal to the world average; it is a little lower in Africa and North America; and it is considerably higher in South America. Antarctica is frozen and Australia is arid, and so they contribute little runoff. Also, since their areas are relatively small, they do not affect the global runoff average significantly. The waters draining the continents have quite different chemistries; those from Europe are very rich in calcium and bicarbonates, whereas those from Africa and South America are not. North American and Asian rivers are somewhat intermediate in their concentrations of these dissolved constituents. Such differences in composition reflect a variety of factors, including runoff, temperature, and relief, but certainly the bulk composition of the continental rocks in contact with these waters and their underground sources play a major role. The surface rocks of Europe are rich in carbonates, and those of South America are not; the latter are dominated by sediments rich in silicate minerals.

The chemistry of groundwater and river runoff is being modified by human activities on a global scale. The natural dissolved riverine input of major constituents to the oceans already has been increased by more than 10 percent because of human activities. In the case of sodium, chlorine, and sulfate, the increases are as high as 30 percent. In the United States alone, total water use is equivalent to one-third of total runoff, with about 2 percent of the water used coming from underground wells. In the southwestern region of the country, water supplies have been tapped heavily and in some areas have been exhausted with no hope of replacement. This extensive use of fresh waters in the United States and throughout the globe makes them particularly susceptible to pollution. Leachates from fertilizers, herbicides, and pesticides are found in some freshwater bodies; toxins or excessive amounts of certain inorganic or organic chemicals are present; radioactive elements have been detected; and some surface water bodies have had their salinities increased dramatically, rendering them useless for human consumption. It is therefore imperative that countries closely monitor the use of freshwater systems and promote their conservation.

Pedosphere

The pedosphere is the outermost layer of the Earth that is composed of soil and subject to soil formation processes. It exists at the interface of the lithosphere, atmosphere, hydrosphere and biosphere. The pedosphere is the skin of the Earth and only develops when there is a dynamic interaction between the atmosphere (air in and above the soil), biosphere (living organisms), lithosphere (unconsolidated regolith and consolidated bedrock) and the hydrosphere (water in, on and below the soil). The pedosphere is the foundation of terrestrial life on Earth.

The pedosphere acts as the mediator of chemical and biogeochemical flux into and out of these respective systems and is made up of gaseous, mineralic, fluid and biologic components. The pedosphere lies within the Critical Zone, a broader interface that includes vegetation, pedosphere, groundwater aquifer systems, regolith and finally ends at some depth in the bedrock where the biosphere and hydrosphere cease to make significant changes to the chemistry at depth. As part of the larger global system, any particular environment in which soil forms is influenced solely by its geographic position on the globe as climatic, geologic, biologic and anthropogenic changes occur with changes in longitude and latitude.

The pedosphere lies below the vegetative cover of the biosphere and above the hydrosphere and lithosphere. The soil forming process (pedogenesis) can begin without the aid of biology but is significantly quickened in the presence of biologic reactions. Soil formation begins with the chemical and/or physical breakdown of minerals to form the initial material that overlies the bedrock substrate. Biology quickens this by secreting acidic compounds (dominantly fulvic acids) that help break rock apart. Particular biologic pioneers are lichen, mosses and seed bearing plants, but many other inorganic reactions take place that diversify the chemical makeup of the early soil layer. Once weathering and decomposition products accumulate, a coherent soil body allows the migration of fluids both vertically and laterally through the soil profile, causing ion exchange between solid, fluid and gaseous phases. As time progresses, the bulk geochemistry of the soil layer will deviate away from the initial composition of the bedrock and will evolve to a chemistry that reflects the type of reactions that take place in the soil.

Lithosphere

The primary conditions for soil development are controlled by the chemical composition of the rock that the soil will eventually be forming on. Rock types that form the base of the soil profile are often either sedimentary (carbonate or siliceous), igneous or metaigneous (metamorphosed igneous rocks) or volcanic and metavolcanic rocks. The rock type and the processes that lead to its exposure at the surface are controlled by the regional geologic setting of the specific area under study, which revolve around the underlying theory of plate tectonics, subsequent deformation, uplift, subsidence and deposition.

Metaigneous and metavolcanic rocks form the largest component of cratons and are high in silica. Igneous and volcanic rocks are also high in silica but with non-metamorphosed rock, weathering becomes faster and the mobilization of ions is more widespread. Rocks high in silica produce silicic acid as a weathering product. There are few rock types that lead to localized enrichment of some of the biologically limiting elements like phosphorus (P) and nitrogen (N). Phosphatic shale ($< 15\%\ P_2O_5$) and phosphorite ($> 15\%\ P_2O_5$) form in anoxic deep water basins that preserve organic material. Greenstone (metabasalt), phyllite and schist release up to 30–50% of the nitrogen pool. Thick successions of carbonate rocks are often deposited on craton margins during sea level rise. The widespread dissolution of carbonate and evaporate minerals leads to elevated levels of Mg^{2+}, HCO_3^-, Sr^{2+}, Na^+, Cl^- and SO_4^{2-} ions in aqueous solution.

Weathering and Dissolution of Minerals

The process of soil formation is dominated by chemical weathering of silicate minerals, aided by acidic products of pioneering plants and organisms as well as carbonic acid inputs from the atmosphere. Carbonic acid is produced in the atmosphere and soil layers through the carbonation reaction.

$$H_2O+CO_2 \rightarrow H^+ + HCO_3^- \rightarrow H_2CO_3$$

This is the dominant form of chemical weathering and aides in the breakdown of carbonate minerals such as calcite and dolomite and silicate minerals such as feldspar. The breakdown of the Na-feldspar, albite, by carbonic acid to form kaolinite clay is as follows:

$$2\ NaAlSi_3O_8 + 2\ H_2CO_3 + 9\ H_2O \rightarrow 2\ Na^+ + 2\ HCO_3^- + 4\ H_4SiO_4 + Al_2Si_2O_5(OH)_4$$

Evidence of this reaction in the field would be elevated levels of bicarbonate (HCO_3^-), sodium and silica ions in the water runoff.

The breakdown of carbonate minerals:

$$CaCO_3 + H_2CO_3 \rightarrow Ca^{2+} + 2\ HCO_3^- \text{ or: } CaCO_3 \rightarrow Ca^{2+} + CO_3^{2-}$$

The further dissolution of carbonic acid (H_2CO_3) and bicarbonate (HCO_3) produces CO_2 gas. Oxidization is also a major contributor to the breakdown of many silicate minerals and formation of secondary minerals (diagenesis) in the early soil profile. Oxidation of olivine ($FeMgSiO_4$) releases Fe, Mg and Si ions. The Mg is soluble in water and is carried in the runoff but the Fe often reacts with oxygen to precipitate Fe_2O_3 (hematite), the oxidized state of iron oxide. Sulfur, a byproduct of decaying organic material will also react with iron to form pyrite (FeS_2) in reducing environments. Pyrite dissolution leads to high pH levels due to elevated H+ ions and further precipitation of Fe_2O_3 ultimately changing the redox conditions of the environment.

Biosphere

Inputs from the biosphere may begin with lichen and other microorganisms that secrete oxalic acid. These microorganisms, associated with the lichen community or independently inhabiting rocks, include a number of blue-green algae, green algae, various fungi, and numerous bacteria. Lichen has long been viewed as the pioneers of soil development as the following statement suggests:

"The initial conversion of rock into soil is carried on by the pioneer lichens and their successors, the mosses, in which the hair-like rhizoids assume the role of roots in breaking down the surface into fine dust"

However, lichens are not necessarily the only pioneering organisms nor the earliest form of soil formation as it has been documented that seed-bearing plants may occupy an area and colonize quicker than lichen. Also, eolian sedimentation can produce high rates of sediment accumulation. Nonetheless, lichen can certainly withstand harsher conditions than most vascular plants and although they have slower colonization rates, do form the dominant group in alpine regions.

Acids released from plant roots include acetic and citric acids. During the decay of organic matter Phenolic acids are released from plant matter and humic and fulvic acids are released by soil microbes. These organic acids speed up chemical weathering by combining with some of the weathering products in a process known as chelation. In the soil profile, the organic acids are often concentrated at the top while carbonic acid plays a larger role towards the bottom or below in the aquifer.

As the soil column develops further into thicker accumulations, larger animals come to inhabit the soil and continue to alter the chemical evolution of their respective niche. Earthworms aerate the soil and convert large amounts of organic matter into rich humus, improving soil fertility. Small burrowing mammals store food, grow young and may hibernate in the pedosphere altering the course of soil evolution. Large mammalian herbivores above ground transport nutrients in form of nitrogen-rich waste and phosphorus-rich antlers while predators leave phosphorus-rich piles of bones on the soil surface, leading the localized enrichment of the soil below.

Redox Conditions in Wetland Soils

Nutrient cycling in lakes and freshwater wetlands depends heavily on redox conditions. Under a few millimeters of water heterotrophic bacteria metabolize and consume oxygen. They therefore deplete the soil of oxygen and create the need for anaerobic respiration. Some anaerobic microbial processes include denitrification, sulfate reduction and methanogenesis and are responsible for the release of N_2 (nitrogen), H_2S (hydrogen sulfide) and CH_4 (methane). Other anaerobic microbial processes are linked to changes in the oxidation state of iron and manganese. As a result of anaerobic decomposition, the soil stores large amounts of organic carbon because decomposition is incomplete.

The redox potential describes which way chemical reactions will proceed in oxygen deficient soils and controls the nutrient cycling in flooded systems. Redox potential, or reduction potential, is used to express the likelihood of an environment to receive electrons and therefore become reduced. For example, if a system already has plenty of electrons (anoxic, organic-rich shale) it is reduced and will likely donate electrons to a part of the system that has a low concentration of electrons, or an oxidized environment, to equilibrate to the chemical gradient. The oxidized environment has high redox potential, whereas the reduced environment has a low redox potential.

The redox potential is controlled by the oxidation state of the chemical species, pH and the amount of oxygen (O_2) there is in the system. The oxidizing environment accepts electrons because of the presence of O_2, which acts as electron acceptors:

$$O_2 + 4\,e^- + 4\,H^+ \rightarrow H_2O$$

This equation will tend to move to the right in acidic conditions which causes higher redox potentials to be found at lower pH levels. Bacteria, heterotrophic organisms, consume oxygen while decomposing organic material which depletes the soils of oxygen, thus increasing the redox potential. In low redox conditions the deposition of ferrous iron (Fe^{2+}) will increase with decreasing decomposition rates, thus preserving organic remains and depositing humus. At high redox potential, the oxidized form of iron, ferric iron (Fe^{3+}), will be deposited commonly as hematite. By using analytical geochemical tools such as x-ray fluorescence (XRF) or inductively coupled mass spectrometry (ICP-MS) the two forms of Fe (Fe^{2+} and Fe^{3+}) can be measured in ancient rocks therefore determining the redox potential for ancient soils.

Such a study was done on Permian through Triassic rocks (300–200 million years old) in Japan and British Columbia. The geologists found hematite throughout the early and middle Permian but began to find the reduced form of iron in pyrite within the ancient soils near the end of the Permian and into the Triassic. This suggests that conditions became less oxygen rich, even anoxic, during the late Permian, which eventually led to the greatest extinction in earth's history, the P-T extinction.

Decomposition in anoxic or reduced soils is also carried out by sulfur-reducing bacteria which, instead of O_2 use SO_4^{2-} as an electron acceptor and produce hydrogen sulfide (H_2S) and carbon dioxide in the process:

$$2\,H^+ + SO_4^{2-} + 2(CH_2O) \rightarrow 2\,CO_2 + H_2S + 2\,H_2O$$

The H_2S gas percolates upwards and reacts with Fe^{2+} and precipitates pyrite, acting as a trap for the toxic H_2S gas. However, H_2S is still a large fraction of emissions from wetland soils.[11] In most freshwater wetlands there is little sulfate (SO_4^{2-}) so methanogenesis becomes the dominant form of decomposition by methanogenic bacteria only when sulfate is depleted. Acetate, a compound that is a byproduct of fermenting cellulose is split by methanogenic bacteria to produce methane (CH_4) and carbon dioxide (CO_2), which are released to the atmosphere. Methane is also released during the reduction of CO_2 by the same bacteria.

Atmosphere

In the pedosphere it is safe to assume that gases are in equilibrium with the atmosphere. Because plant roots and soil microbes release CO_2 to the soil, the concentration of bicarbonate(HCO_3) in soil waters is much greater than that in equilibrium with the atmosphere, the high concentration of CO_2 and the occurrence of metals in soil solutions results in lower pH levels in the soil. Gases that escape from the pedosphere to the atmosphere include the gaseous byproducts of carbonate dissolution, decomposition, redox reactions and microbial photosynthesis. The main inputs from the atmosphere are aeolian sedimentation, rainfall and gas diffusion. Eolian sedimentation includes anything that can be entrained by wind or that stays suspended, seemingly indefinitely, in air and includes a wide variety of aerosol particles, biological particles like pollen and dust to pure quartz sand. Nitrogen is the most abundant constituent in rain (after water), as water vapor utilizes aerosol particles to nucleate rain droplets.

Soil in Forests

Soil is well developed in the forest as suggested by the thick humus layers, rich diversity of large trees and animals that live there. In forests, precipitation exceeds evapotranspiration which results in an excess of water that percolates downward through the soil layers. Slow rates of decomposition leads to large amounts of fulvic acid, greatly enhancing chemical weathering. The downward percolation, in conjunction with chemical weathering leaches magnesium (Mg), iron (Fe), and aluminium (Al) from the soil and transports them downward, a process known as podzolization. This process leads to marked contrasts in the appearance and chemistry of the soil layers.

Soil in the Tropics

Tropical forests (rainforests) receive more insolation and rainfall over longer growing seasons than any other environment on earth. With these elevated temperatures, insolation and rainfall, biomass is extremely productive leading to the production of as much as 800 grams of carbon per square meter per year. Higher temperatures and larger amounts of water contribute to higher rates of chemical weathering. Increased rates of decomposition cause smaller amounts of fulvic acid to percolate and leach metals from the zone of active weathering. Thus, in stark contrast to soil in forests, tropical forests have little to no podzolization and therefore do not have marked visual and chemical contrasts with the soil layers. Instead, the mobile metals Mg, Fe and Al are precipitated as oxide minerals giving the soil a rusty red color.

Soil in Grasslands and Deserts

Precipitation in grasslands is equal to or less than evapotranspiration and causes soil development to operate in relative drought. Leaching and migration of weathering products is therefore decreased. Large amounts of evaporation causes buildup of calcium (Ca) and other large cations flocculate clay minerals and fulvic acids in the upper soil profile. Impermeable clay limits downward percolation of water and fulvic acids, reducing chemical weathering and podzolization. The depth to the maximum concentration of clay increases in areas of increased precipitation and leaching. When leaching is decreased, the Ca precipitates as calcite ($CaCO_3$) in the lower soil levels, a layer known as caliche.

Deserts behave similarly to grasslands but operate in constant drought as precipitation is less than evapotranspiration. Chemical weathering proceeds more slowly than in grasslands and beneath the caliche layer may be a layer of gypsum and halite. To study soils in deserts, pedologists have used the concept of chronosequences to relate timing and development of the soil layers. It has been shown that P is leached very quickly from the system and therefore decreases with increasing age. Furthermore, carbon buildup in the soils is decreased due to slower decomposition rates. As a result, the rates of carbon circulation in the biogeochemical cycle is decreased.

Earth Sciences

Earth sciences is the field of study concerned with the solid Earth, its waters, and the air that envelops it. Included are the geologic, hydrologic, and atmospheric sciences.

The broad aim of the Earth sciences is to understand the present features and the past evolution of the Earth and to use this knowledge, where appropriate, for the benefit of humankind. Thus the basic concerns of the Earth scientist are to observe, describe, and classify all the features of the Earth, whether characteristic or not, to generate hypotheses with which to explain their presence and their development, and to devise means of checking opposing ideas for their relative validity. In this way the most plausible, acceptable, and long-lasting ideas are developed.

The physical environment in which humans live includes not only the immediate surface of the solid Earth, but also the ground beneath it and the water and air above it. Early man was more involved with the practicalities of life than with theories, and thus his survival depended on his ability to obtain metals from the ground to produce, for example, alloys, such as bronze from copper and tin, for tools and armour, to find adequate water supplies for establishing dwelling sites, and to forecast the weather, which had a far greater bearing on human life in earlier times than it has today. Such situations represent the foundations of the three principal component disciplines of the modern Earth sciences.

The rapid development of science as a whole over the past century and a half has given rise to an immense number of specializations and subdisciplines, with the result that the modern Earth scientist, perhaps unfortunately, tends to know a great deal about a very small area of study but only a little about most other aspects of the entire field. Only when one is aware of the marvelous complexity of the Earth sciences and yet can understand the breakdown of the component disciplines is one in a position to select those parts of the subject that are of greatest personal interest.

It is worth emphasizing two important features that the three divisions of the Earth sciences have in common. First is the inaccessibility of many of the objects of study. Many rocks, as well as water and oil reservoirs, are at great depths in the Earth, while air masses circulate at vast heights above it. Thus the Earth scientist has to have a good three-dimensional perspective. Second, there is the fourth dimension: time. The Earth scientist is responsible for working out how the Earth evolved over millions of years. For example, what were the physical and chemical conditions operating on the Earth and the Moon 3.5 billion years ago? How did the oceans form, and how did their chemical composition change with time? How has the atmosphere developed? And finally, how did life on Earth begin, and from what did man evolve?

Today the Earth sciences are divided into many disciplines, which are themselves divisible into six groups:

1. Those subjects that deal with the water and air at or above the solid surface of the Earth. These include the study of the water on and within the ground (hydrology), the glaciers and ice caps (glaciology), the oceans (oceanography), the atmosphere and its phenomena (meteorology), and the world's climates (climatology).

2. Disciplines concerned with the physical-chemical makeup of the solid Earth, which include the study of minerals (mineralogy), the three main groups of rocks (igneous, sedimentary, and metamorphic petrology), the chemistry of rocks (geochemistry), the structures in rocks (structural geology), and the physical properties of rocks at the Earth's surface and in its interior (geophysics).

3. The study of landforms (geomorphology), which is concerned with the description of the features of the present terrestrial surface and an analysis of the processes that gave rise to them.

4. Disciplines concerned with the geologic history of the Earth, including the study of fossils and the fossil record (paleontology), the development of sedimentary strata deposited typically over millions of years (stratigraphy), and the isotopic chemistry and age dating of rocks (geochronology).

5. Applied Earth sciences dealing with current practical applications beneficial to society. These include the study of fossil fuels (oil, natural gas, and coal); oil reservoirs; mineral deposits; geothermal energy for electricity and heating; the structure and composition of bedrock for the location of bridges, nuclear reactors, roads, dams, and skyscrapers and other buildings; hazards involving rock and mud avalanches, volcanic eruptions, earthquakes, and the collapse of tunnels; and coastal, cliff, and soil erosion.

6. The study of the rock record on the Moon and the planets and their satellites (astrogeology). This field includes the investigation of relevant terrestrial features—namely, tektites (glassy objects resulting from meteorite impacts) and astroblemes (meteorite craters).

With such intergradational boundaries between the divisions of the Earth sciences (which, on a broader scale, also intergrade with physics, chemistry, biology, mathematics, and certain branches of engineering), researchers today must be versatile in their approach to problems. Hence, an important aspect of training within the Earth sciences is an appreciation of their multidisciplinary nature.

Origins in Prehistoric Times

The origins of the Earth sciences lie in the myths and legends of the distant past. The creation story, which can be traced to a Babylonian epic of the 22nd century BCE and which is told in the first chapter of Genesis, has proved most influential. The story is cast in the form of Earth history and thus was readily accepted as an embodiment of scientific as well as of theological truth.

Earth scientists later made innumerable observations of natural phenomena and interpreted them in an increasingly multidisciplinary manner. The Earth sciences, however, were slow to develop largely because the progress of science was constrained by whatever society would tolerate or support at any one time.

Antiquity

Geologic Sciences

The oldest known treatise on rocks and minerals is the *De lapidibus* ("On Stones") of the Greek philosopher Theophrastus. Written probably in the early years of the 3rd century, this work remained the best study of mineral substances for almost 2,000 years. Although reference is made to some 70 different materials, the work is more an effort at classification than systematic description.

In early Chinese writings on mineralogy, stones and rocks were distinguished from metals and alloys, and further distinctions were made on the basis of colour and other physical properties. The speculations of Zheng Sixiao on the origin of ore deposits were more advanced than those of his contemporaries in Europe. In brief, his theory was that ore is deposited from groundwater circulating in subsurface fissures.

Ancient accounts of earthquakes and volcanic eruptions are sometimes valuable as historical records but tell little about the causes of these events. Aristotle and Strabo held that volcanic explosions and earthquakes alike are caused by spasmodic motions of hot winds that move underground and occasionally burst forth in volcanic activity attended by Earth tremors. Classical and medieval ideas on earthquakes and volcanoes were brought together in William Caxton's *Mirrour of the World*. Earthquakes are here again related to movements of subterranean fluids. Streams of water in the Earth compress the air in hidden caverns. If the roofs of the caverns are weak, they rupture, causing cities and castles to fall into the chasms; if strong, they merely tremble and shake from the heaving by the wind below. Volcanic action follows if the outburst of wind and water from the depths is accompanied by fire and brimstone from hell.

The Chinese have the distinction of keeping the most faithful records of earthquakes and of inventing the first instrument capable of detecting them. Records of the dates on which major quakes rocked China date to 780 BCE. To detect quakes at a distance, the mathematician, astronomer, and geographer Zhang Heng invented what has been called the first seismograph.

Knowledge of Earth History

The occurrence of seashells embedded in the hard rocks of high mountains aroused the curiosity of early naturalists and eventually set off a controversy on the origin of fossils that continued through the 17th century. Xenophanes of Colophon was credited by later writers with observing that seashells occur "in the midst of earth and in mountains." He is said to have believed that these relics originated during a catastrophic event that caused the earth to be mixed with the sea and then to settle, burying organisms in the drying mud. For these views Xenophanes is sometimes called the father of paleontology.

Petrified wood was described by Chinese scholars as early as the 9th century CE, and about 1080 Shen Gua cited fossilized plants as evidence for change in climate. Other kinds of fossils that attracted the attention of early Chinese writers include spiriferoid brachiopods ("stone swallows"), cephalopods, crabs, and the bones and teeth of reptiles, birds, and mammals. Although these objects were commonly collected simply as curiosities or for medicinal purposes, Shen Gua recognized marine invertebrate fossils for what they are and for what they imply historically. Observing seashells in strata of the Taihang Mountains, he concluded that this region, though now far from the sea, must once have been a shore.

Knowledge of Landforms and of Land-sea Relations

Changes in the landscape and in the position of land and sea related to erosion and deposition by streams were recognized by some early writers. The Greek historian Herodotus correctly concluded that the northward bulge of Egypt into the Mediterranean is caused by the deposition of mud carried by the Nile.

The early Chinese writers were not outdone by the Romans and Greeks in their appreciation of changes wrought by erosion. In the *Jinshu* ("History of the Jin Dynasty"), it is said of Du Yu that when he ordered monumental stelae to be carved with the records of his successes, he had one buried at the foot of a mountain and the other erected on top. He predicted that in time they would likely change their relative positions, because the high hills will become valleys and the deep valleys will become hills.

Aristotle guessed that changes in the position of land and sea might be cyclical in character, thus reflecting some sort of natural order. If the rivers of a moist region should build deltas at their mouths, he reasoned, seawater would be displaced and the level of the sea would rise to cover some adjacentdry region. A reversal of climatic conditions might cause the sea to return to the area from which it had previously been displaced and retreat from the area previously inundated. The idea of a cyclical interchange between land and sea was elaborated in the Discourses of the Brothers of Purity, a classic Arabic work written between 941 and 982 CE by an anonymous group of scholars at Basra (Iraq). The rocks of the lands disintegrate and rivers carry their wastage to the sea, where waves and currents spread it over the seafloor. There the layers of sediment accumulate one above the other, harden, and, in the course of time, rise from the bottom of the sea to form new continents. Then the process of disintegration and leveling begins again.

Hydrologic and Atmospheric Sciences

The only substance known to the ancient philosophers in its solid, liquid, and gaseous states, water is prominently featured in early theories about the origin and operations of the Earth. Thales of Miletus(*c.* 624–*c.* 545 BCE) is credited with a belief that water is the essential substance of the Earth, and Anaximander of Miletus (*c.* 610–545 BCE) held that water was probably the source of life. In the system proposed by Empedocles of Agrigentum (*c.* 490–430 BCE), water shared the primacy Thales had given it with three other elements: fire, air, and earth. The doctrine of the four earthly elements was later embodied in the universal system of Aristotle and thereby influenced Western scientific thought until late in the 17th century.

Knowledge of the Hydrologic Cycle

The idea that the waters of the Earth undergo cyclical motions, changing from seawater to vapour to precipitation and then flowing back to the ocean, is probably older than any of the surviving texts that hint at or frame it explicitly.

The idea of the hydrological cycle developed independently in China as early as the 4th century BCE and was explicitly stated in the Lüshi chunqiu ("The Spring and Autumn [Annals] of Mr. Lü"), written in the 3rd century BCE. A circulatory system of a different kind, involving movements of water on a large scale within the Earth, was envisioned by Plato (*c.* 428–348/347 BCE). In one of his two explanations for the origin of rivers and springs, he described the Earth as perforated by passages connecting with Tartarus, a vast subterranean reservoir.

A coherent theory of precipitation is found in the writings of Aristotle. Moisture on the Earth is changed to airy vapour by heat from above. Because it is the nature of heat to rise, the heat in the vapour carries it aloft. When the heat begins to leave the vapour, the vapour turns to water. The formation of water from air produces clouds. Heat remaining in the clouds is further opposed by the cold inherent in the water and is driven away. The cold presses the particles of the cloud closer together, restoring in them the true nature of the element water. Water naturally moves downward, and so it falls from the cloud as raindrops. Snow falls from clouds that have frozen.

In Aristotle's system the four earthly elements were not stable but could change into one another. If air can change to water in the sky, it should also be able to change into water underground.

The Origin of the Nile

Of all the rivers known to the ancients, the Nile was most puzzling with regard to its sources of water. Not only does this river maintain its course up the length of Egypt through a virtually rainless desert, but it rises regularly in flood once each year.

Speculations on the strange behaviour of the Nile were many, varied, and mostly wrong. Thales suggested that the strong winds that blow southward over the delta in summertime hold back the flow of the river and cause the waters to rise upstream in flood. Oenopides of Chios (flourished *c.* 475 BCE) thought that heat stored in the ground during the winter dries up the underground veins of water so that the river shrinks. In the summer the heat disappears, and water flows up into the river, causing floods. In the view of Diogenes of Apollonia (flourished *c.* 435 BCE), the Sun controls the regimen of the stream. The idea that the Nile waters connect with the sea is an old one, tracing back to the geographic concepts of Hecataeus of Miletus (*c.* 520 BCE). Reasonable explanations related the discharge of the Nile to precipitation in the headwater regions, as snow (Anaxagoras of

Clazomenae, *c.* 500–428 BCE) or from rain that filled lakes supposed to have fed the river (Democritus of Abdera, *c.*460–*c.* 357 BCE). Eratosthenes (*c.* 276–194 BCE), who had prepared a map of the Nile valley southward to the latitude of modern Khartoum, anticipated the correct answer when he reported that heavy rains had been observed to fall in the upper reaches of the river and that these were sufficient to account for the flooding.

Knowledge of the Tides

The tides of the Mediterranean, being inconspicuous in most places, attracted little notice from Greek and Roman naturalists. Poseidonius (135–50 BCE) first correlated variations in the tides with phases of the Moon. By contrast, the tides along the eastern shores of Asia generally have a considerable range and were the subject of close observation and much speculation among the Chinese. In particular, the tidal bore on the Qiantang River near Hangzhou attracted early attention; with its front ranging up to 3.7 metres in height, this bore is one of the largest in the world. As early as the 2nd century BCE, the Chinese had recognized a connection between tides and tidal bores and the lunar cycle.

Prospecting for Groundwater

Although the origin of the water in the Earth that seeps or springs from the ground was long the subject of much fanciful speculation, the arts of finding and managing groundwater were already highly developed in the 8th century BCE. The construction of long, hand-dug underground aqueducts (*qanāts*) in Armenia and Persia represents one of the great hydrologic achievements of the ancient world. After some 3,000 years *qanāts* are still a major source of water in Iran.

In the 1st century BCE, Vitruvius (Marcus Vitruvius Pollio), a Roman architect and engineer, described methods of prospecting for groundwater in his De architectura *libri decem* (*The Architecture of Marcus Vitruvius Pollio, in Ten Books*). To locate places where wells should be dug, he recommended looking for spots where mist rises in early morning. More significantly, Vitruvius had learned to associate different quantities and qualities of groundwater with different kinds of rocks and topographic situations.

After the inspired beginnings of the ancient Greeks, Romans, Chinese, and Arabs, little or no new information was collected, and no new ideas were produced throughout the Middle Ages, appropriately called the Dark Ages. It was not until the Renaissance in the early 16th century that the Earth sciences began to develop again.

The 16th–18th Centuries

Geologic Sciences

Ore Deposits and Mineralogy

A common belief among alchemists of the 16th and 17th centuries held that metalliferous deposits were generated by heat emanating from the centre of the Earth but activated by the heavenly bodies.

The German scientist Georgius Agricolahas with much justification been called the father of mineralogy.

Of his seven geologic books, De natura fossilium (1546; "On Natural Fossils") contains his major contributions to mineralogy and, in fact, has been called the first textbook on that subject. In Agricola's time and well into the 19th century, "fossil" was a term that could be applied to any object dug from the Earth. Thus Agricola's classification of fossils provided pigeonholes for organic remains, such as ammonites, and for rocks of various kinds in addition to minerals. Individual kinds of minerals, their associations and manners of occurrence, are described in detail, many for the first time.

With the birth of analytical chemistry toward the latter part of the 18th century, the classification of minerals on the basis of their composition at last became possible. The German geologist Abraham Gottlob Werner was one of those who favoured a chemical classification in preference to a "natural history" classification based on external appearances. His list of several classifications, published posthumously, recognized 317 different substances ordered in four classes.

Paleontology and Stratigraphy

During the 17th century the guiding principles of paleontology and historical geology began to emerge in the work of a few individuals. Nicolaus Steno, a Danish scientist and theologian, presented carefully reasoned arguments favouring the organic origin of what are now called fossils. Also, he elucidated three principles that made possible the reconstruction of certain kinds of geologic events in a chronological order. In his *Canis carcariae dissectum caput* (1667; "Dissected Head of a Dog Shark"), he concluded that large tongue-shaped objects found in the strata of Malta were the teeth of sharks, whose remains were buried beneath the seafloor and later raised out of the water to their present sites. This excursion into paleontology led Steno to confront a broader question. How can one solid body, such as a shark's tooth, become embedded in another solid body, such as a layer of rock? He published his answers in 1669 in a paper titled "De solido intra naturaliter contento dissertationis" ("A Preliminary Discourse Concerning a Solid Body Enclosed by Processes of Nature Within a Solid"). Steno cited evidence to show that when the hard parts of an organism are covered with sediment, it is they and not the aggregates of sediment that are firm. Consolidation of the sediment into rock may come later, and, if so, the original solid fossil becomes encased in solid rock. He recognized that sediments settle from fluids layer by layer to form strata that are originally continuous and nearly horizontal. His principle of superposition of strata states that in a sequence of strata, as originally laid down, any stratum is younger than the one on which it rests and older than the one that rests upon it.

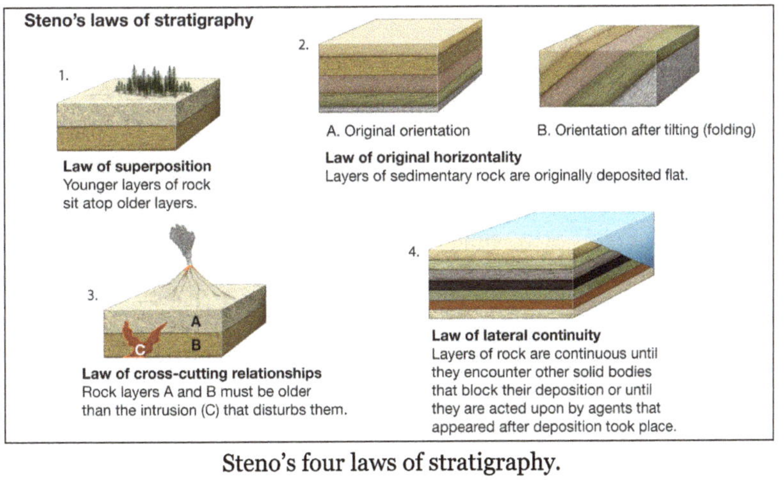

Steno's four laws of stratigraphy.

In 1667 and 1668, the English physicist Robert Hooke read papers before the Royal Society in which he expressed many of the ideas contained in Steno's works. Hooke argued for the organic nature of fossils. Elevation of beds containing marine fossils to mountainous heights he attributed to the work of earthquakes. Streams attacking these elevated tracts wear down the hills, fill depressions with sediment, and thus level out irregularities of the landscape.

Earth History According to Werner and James Hutton

The two major theories of the 18th century were the Neptunian and the Plutonian. The Neptunists, led by Werner and his students, maintained that the Earth was originally covered by a turbid ocean. The first sediments deposited over the irregular floor of this universal ocean formed the granite and other crystalline rocks. Then as the ocean began to subside, "Stratified" rocks were laid down in succession. The "Volcanic" rocks were the youngest; Neptunists took small account of volcanism and thought that lava was formed by the burning of coal deposits underground.

The Scottish scientist James Hutton, leader of the Plutonists, viewed the Earth as a dynamic body that functions as a heat machine. Streams wear down the continents and deposit their waste in the sea. Subterranean heat causes the outer part of the Earth to expand in places, uplifting the compacted marine sediments to form new continents. Hutton recognized that granite is an intrusive igneous rockand not a primitive sediment as the Neptunists claimed. Intrusive sills and dikes of igneous rockprovide evidence for the driving force of subterranean heat. Hutton viewed great angular unconformities separating sedimentary sequences as evidence for past cycles of sedimentation, uplift, and erosion. His Theory of the Earth, published as an essay in 1788, was expanded to a two-volume work in 1795. John Playfair, a professor of natural philosophy, defended Hutton against the counterattacks of the Neptunists, and his Illustrations of the Huttonian Theory (1802) is the clearest contemporary account of Plutonist theory.

Hydrologic Sciences

The idea that there is a circulatory system within the Earth, by which seawater is conveyed to mountaintops and there discharged, persisted until early in the 18th century. Two questions left unresolved by this theory were acknowledged even by its advocates. How is seawater forced uphill? How is the salt lost in the process?

The Rise of Subterranean Water

René Descartes supposed that the seawater diffused through subterranean channels into large caverns below the tops of mountains. The Jesuit philosopher Athanasius Kircherin his Mundus subterraneus (1664; "Subterranean World") suggested that the tides pump seawater through hidden channels to points of outlet at springs. To explain the rise of subterranean water beneath mountains, the chemist Robert Plot appealed to the pressure of air, which forces water up the insides of mountains. The idea of a great subterranean sea connecting with the ocean and supplying it with water together with all springs and rivers was resurrected in 1695 in John Woodward's *Essay* Towards a Natural History of the Earth and Terrestrial Bodies.

The French Huguenot Bernard Palissy maintained, to the contrary, that rainfall is the sole source of rivers and springs. In his *Discours admirables* (1580; Admirable Discourses) he described how rainwater falling on mountains enters cracks in the ground and flows down along these until, diverted by some obstruction, it flows out on the surface as springs. Palissy scorned the idea that seawater courses in veins to the tops of mountains. For this to be true, sea level would have to be higher than mountaintops—an impossibility. In his *Discours* Palissy suggested that water would rise above the level at which it was first encountered in a well provided the source of the groundwater came from a place higher than the bottom of the well. This is an early reference to conditions essential to the occurrence of artesian water, a popular subject among Italian hydrologists of the 17th and 18th centuries.

In the latter part of the 17th century, Pierre Perrault and Edmé Mariotte conducted hydrologic investigations in the basin of the Seine River that established that the local annual precipitation was more than ample to account for the annual runoff.

Evaporation from the Sea

The question remained as to whether the amount of water evaporated from the sea is sufficient to account for the precipitation that feeds the streams. The English astronomer-mathematician Edmond Halley measured the rate of evaporation from pans of water exposed to the air during hot summer days. Assuming that this same rate would obtain for the Mediterranean, Halley calculated that some 5.28 billion tons of water are evaporated from this sea during a summer day. Assuming further that each of the nine major rivers flowing into the Mediterranean has a daily discharge 10 times that of the Thames, he calculated that a daily inflow of fresh water back into that sea would be 1.827 billion tons, only slightly more than a third of the amount lost by evaporation. Halley went on to explain what happens to the remainder. A part falls back into the sea as rain before it reaches land. Another part is taken up by plants.

In the course of the hydrologic cycle, Halley reasoned, the rivers constantly bring salt into the sea in solution, but the salt is left behind when seawater evaporates to replenish the streams with rainwater. Thus the sea must be growing steadily saltier.

Atmospheric Sciences

Water Vapour in the Atmosphere

After 1760 the analytical chemists at last demonstrated that water and air are not the same substance in different guises. Long before this development, however, investigators had begun to draw a distinction between water vapour and air. Otto von Guericke, a German physicist and engineer, produced artificial clouds by releasing air from one flask into another one from which the air had been evacuated. A fog then formed in the unevacuated flask. Guericke concluded that air cannot be turned into water, though moisture can enter the air and later be condensed into water. Guericke's experiments, however, did not answer the question as to how water enters the atmosphere as vapour. In "Les Météores"("Meteorology," an essay published in the book *Discours de la methode*in 1637), Descartes envisioned water as composed of minute particles that were elongate, smooth, and separated by a highly rarified "subtle matter."

The same uncertainty as to how water gets into the air surrounded the question as to how it remains suspended as clouds. A popular view in the 18th century was that clouds are made of countless tiny bubbles that float in air. Guericke had suggested that the fine particles in his artificial clouds were bubbles. Other observers professed to have seen bubble-shaped particles of water vapour rising from warm water or hot coffee.

Pressure, Temperature and Atmospheric Circulation

If clouds are essentially multicompartmented balloons, their motions could be explained by the movements of winds blowing on them. Descartes suggested that the winds might blow upward as well as laterally, causing the clouds to rise or at least preventing them from descending. In 1749 Benjamin Franklin explained updrafts of air as due to local heating of the atmosphere by the Sun. Sixteen years later the Swiss-German mathematical physicist Johann Heinrich Lambert described the conditions necessary for the initiation of convection currents in the atmosphere. He reasoned that rising warm air flows into bordering areas of cooler air, increasing their downward pressure and causing their lower layers to flow into ascending currents, thus producing circulation.

The fact that Lambert could appeal to changes in air pressure to explain circulation reflects an important change from the view still current in the late 16th century that air is weightless. This misconception was corrected after 1643 with the invention of the mercury barometer. It was soon discovered that the height of the barometer varied with the weather, usually standing at its highest during clear weather and falling to the lowest on rainy days.

Toward the end of the 18th century it was beginning to be understood that variations in the barometer must be related to the general motion and circulation of the atmosphere. That these variations could not be due solely to changes in humidity was the conclusion of the Swiss scientist Horace Bénédict de Saussure in his *Essais sur l'hygrométrie* (1783; "Essay on Hygrometry"). From experiments with changes of water vapour and pressure in air enclosed in a glass globe, Saussure concluded that changes in temperature must be immediately responsible for variations of the barometer and that these in turn must be related to the movement of air from one place to another.

The 19th Century

Geologic Sciences

Crystallography and the Classification of Minerals and Rocks

The French scientist René-Just Häuy, whose treatises on mineralogy and crystallography appeared in 1801 and 1822, respectively, has been credited with advancing mineralogy to the status of a science and with establishing the science of crystallography. From his studies of the geometric relationships between planes of cleavage, he concluded that the ultimate particles forming a given species of mineral have the same shape and that variations in crystal habit reflect differences in the ways identical molecules are put together. In 1814 Jöns Jacob Berzelius of Sweden published a system of mineralogy offering a comprehensive classification of minerals based on their chemistry. Berzelius recognized silica as an acid and introduced into mineralogy the group known as silicates. At mid-century the American geologist James Dwight Dana's System of Mineralogy, in its third edition, was reorganized around a chemical classification, which thereafter became standard for handbooks.

The development of the polarizing microscope and the technique for grinding sections of rocks so thin as to be virtually transparent came in 1827 from studies of fossilized wood by William Nicol. In 1849 Clifton Sorby showed that minerals viewed in thin section could be identified by their optical properties, and soon afterward improved classifications of rocks were made on the basis of their mineralogic composition. The German geologist Ferdinand Zirkel's *Mikroscopische Beschaffenheit der Mineralien und Gesteine* (1873; "The Microscopic Nature of Minerals and Rocks") contains one of the first mineralogic classifications of rocks and marks the emergence of microscopic petrography as an established branch of science.

William Smith and Faunal Succession

In 1683 the zoologist Martin Lister proposed to the Royal Society that a new sort of map be drawn showing the areal distribution of the different kinds of British "soiles" (vegetable soils and underlying bedrock). The work proposed by Lister was not accomplished until 132 years later, when William Smith published his Geologic Map of England and Wales with Part of Scotland. A self-educated surveyor and engineer, Smith had the habit of collecting fossils and making careful note of the strata that contained them. He discovered that the different stratified formations in England contain distinctive assemblages of fossils. His map, reproduced on a scale of five miles to the inch, showed 20 different rock units, to which Smith applied local names in common use—e.g., London Clay and Purbeck Beds. In 1816 Smith published a companion work, *Strata Identified by Organized Fossils*, in which the organic remains characteristic of each of his rock units were illustrated. His generalization that each formation is "possessed of properties peculiar to itself [and] has the same organized fossils throughout its course" is the first clear statement of the principle of faunal sequence, which is the basis for worldwide correlation of fossiliferous strata into a coherent system. Smith thus demonstrated two kinds of order in nature: order in the spatial arrangement of rock units and order in the succession of ancient forms of life.

Smith's principle of faunal sequence was another way of saying that there are discontinuities in the sequences of fossilized plants and animals. These discontinuities were interpreted in two ways: as indicators of episodic destruction of life or as evidence for the incompleteness of the fossil record. Baron Georges Cuvier of France was one of the more distinguished members of a large group of naturalists who believed that paleontological discontinuities bore witness to sudden and widespread catastrophes. Cuvier's skill at comparative anatomy enabled him to reconstruct from fragmentary remains the skeletons of large vertebrate animals found at different levels in the Cenozoic sequence of northern France. From these studies he discovered that the fossils in all but the youngest deposits belong to species now extinct. Moreover, these extinct species have definite ranges up and down in the stratigraphic column. Cuvier inferred that the successive extinctions were the result of convulsions that caused the strata of the continents to be dislocated and folded and the seas to sweep across the continents and just as suddenly subside.

Charles Lyell and Uniformitarianism

In opposition to the catastrophist school of thought, the British geologist Charles Lyell proposed a uniformitarian interpretation of geologic history in his Principles of Geology. His system was based on two propositions: the causes of geologic change operating include all the causes that have acted from the earliest time; and these causes have always operated at the same average levels of energy. These two propositions add up to a "steady-state" theory of the Earth. Changes in climate have fluctuated around a mean, reflecting changes in the position of land and sea. Progress through time in the organic world is likewise an illusion, the effect of an imperfect paleontological record. The main part of the *Principles* was devoted less to theory than to procedures for inferring events from rocks; and for Lyell's clear exposition of methodology his work was highly regarded throughout its many editions, long after the author himself had abandoned antiprogressivist views on the development of life.

Louis Agassiz and the Ice Age

Huge boulders of granite resting upon limestone of the Jura Mountains were subjects of controversy during the 18th and early 19th centuries. Saussure described these in 1779 and called them erratics. He concluded that they had been swept to their present positions by torrents of water. Saussure's interpretation was in accord with the tenets of diluvial geologists, who interpreted erratics and sheets of unstratified sediment (till or drift) spread over the northern parts of Europe and North America as the work of the "Deluge."

In 1837 the Swiss zoologist and paleontologist Louis Agassiz delivered a startling address before the Helvetian Society, proposing that, during a geologically recent stage of refrigeration, glacial ice had covered Eurasia from the North Pole to the shores of the Mediterranean and Caspian seas.

Wherever erratics, till, and striated pavements of rock occur, sure evidence of this recent catastrophe exists. The reception accorded this address was glacial, too, and Alexander von Humboldt advised Agassiz to return to his fossil fishes. Instead, he began intensive field studies and in 1840 published his *Études sur les glaciers* ("Studies of Glaciers"), demonstrating that Alpine glaciers had been far more extensive in the past. That same year he visited the British Isles in the company of Buckland and extended the glacial doctrine to Scotland, northern England, and Ireland. In 1846 he carried his campaign to North America and there found additional evidence for an ice age.

Geologic Time and the Age of the Earth

By mid-century the fossiliferous strata of Europe had been grouped into systems arrayed in chronological order. The stratigraphic column, a composite of these systems, was pieced together from exposures in different regions by application of the principles of superposition and faunal sequence. Time elapsed during the formation of a system became known as a period, and the periods were grouped into eras: the Paleozoic (Cambrian through Permian periods), Mesozoic (Triassic, Jurassic, and Cretaceous periods), and Cenozoic (Paleogene, Neogene, and Quaternary periods).

Charles Darwin's Origin of Species offered a theoretical explanation for the empirical principle of faunal sequence. The fossils of the successive systems are different not only because parts of the stratigraphic record are missing but also because most species have lost in their struggles for survival and also because those that do survive evolve into new forms over time. Darwin borrowed two ideas from Lyell and the uniformitarians: the idea that geologic time is virtually without limit and the idea that a sequence of minute changes integrated over long periods of time produce remarkable changes in natural entities.

The evolutionists and the historical geologists were embarrassed when, beginning in 1864, William Thomson (later Lord Kelvin) attacked the steady-state theory of the Earth and placed numerical strictures on the length of geologic time. The Earth might function as a heat machine, but it could not also be a perpetual motion machine. Assuming that the Earth was originally molten, Thomson calculated that not less than 20 million and not more than 400 million years could have passed since the Earth first became a solid body. Other physicists of note put even narrower limits on the Earth's age ranging down to 15 million or 20 million years. All these calculations, however, were based on the common assumption, not always explicitly stated, that the Earth's substance is inert and hence incapable of generating new heat. Shortly before the end of the century this assumption was negated by the discovery of radioactive elements that disintegrate spontaneously and release heat to the Earth in the process.

Concepts of Landform Evolution

The scientific exploration of the American West following the end of the Civil War yielded much new information on the sculpture of the landscape by streams. John Wesley Powellin his reports on the Colorado River and Uinta Mountains explained how streams may come to flow across mountain ranges rather than detour around them. The Green River does not follow some structural crack in its gorge across the Uinta Mountains; instead it has cut its canyon as the mountain range was slowly bowed up. Given enough time, streams will erode their drainage basins to plains approaching sea level as a base. Grove Karl Gilbert's *Report on the Geology of the Henry Mountains* offered a detailed analysis of fluvial processes. According to Gilbert all streams work toward a graded condition, a state of dynamic equilibrium that is attained when the net effect of the flowing water is neither erosion of the bed nor deposition of sediment, when the landscape reflects a balance between the resistance of the rocks to erosion and the processes that are operative upon them. After 1884 William Morris Davis developed the concept of the geographical cycle, during which elevated regions pass through successive stages of dissection and denudation characterized as youthful, mature, and old. Youthful landscapes have broad divides and narrow valleys. With further denudation the original surface on which the streams began their work is reduced to ridgetops. Finally in the stage of old age, the region is reduced to a nearly featureless plain near sea level or its inland projection. Uplift of the region in any stage of this evolution will activate a new cycle. Davis's views dominated geomorphic thought until well into the 20th century, when quantitative approaches resulted in the rediscovery of Gilbert's ideas.

Gravity, Isostasy and the Earth's Figure

Discoveries of regional anomalies in the Earth's gravity led to the realization that high mountain ranges have underlying deficiencies in mass about equal to the apparent surface loads represented by the mountains themselves. In the 18th century the French scientist Pierre Bouguer had observed that the deflections of the pendulum in Peru are much less than they should be if the Andes represent a load perched on top of the Earth's crust. Similar anomalies were later found to obtain along the Himalayan front. To explain these anomalies it was necessary to assume that beneath some depth within the Earth pressures are hydrostatic (equal on all sides). If excess loads are placed upon the crust, as by addition of a continental ice cap, the crust will sink to compensate for the additional mass and will rise again when the load is removed. The tendency toward general equilibrium maintained through vertical movements of the Earth's outer layers was called isostasy in 1899 by Clarence Edward Dutton of the United States.

Evidence for substantial vertical movements of the crust was supplied by studies of regional stratigraphy. In 1883 another American geologist, James Hall, had demonstrated that Paleozoic rocks of the folded Appalachians were several times as thick as sequences of the same age in the plateaus and plains to the west. It was his conclusion that the folded strata in the mountains must have accumulated in a linear submarine trough that filled with sediment as it subsided. Downward crustal flexures of this magnitude came to be called geosynclines.

Hydrologic Sciences

Darcy's Law

Quantitative studies of the movement of water in streams and aquifers led to the formulation of mathematical statements relating discharge to other factors. Henri-Philibert-Gaspard Darcy, a French hydraulic engineer, was the first to state clearly a law describing the flow of groundwater. Darcy's experiments, reported in 1856, were based on the ideas that an aquifer is analogous to a main line connecting two reservoirs at different levels and that an artesian well is like a pipe drawing water from a main line under pressure. His investigations of flow through stratified beds of sand led him to conclude that the rate of flow is directly proportional to the energy loss and inversely proportional to the length of the path of flow. Another French engineer, Arsène-Jules-Étienne-Juvénal Dupuit, extended Darcy's work and developed equations for underground flow toward a well, for the recharge of aquifers, and for the discharge of artesian wells. Philip Forchheimer, an Austrian hydrologist, introduced the theory of functions of a complex variable to analyze the flow by gravity of underground water toward wells and developed equations for determining the critical distance between a river and a well beyond which water from the river will not move into the well.

Surface Water Discharge

A complicated empirical formula for the discharge of streams resulted from the studies of Andrew Atkinson Humphreys and Henry Larcom Abbot in the course of the Mississippi Delta Survey of 1851–60. Their formula contained no term for roughness of channel and on this and other grounds was later found to be inapplicable to the rapidly flowing streams of mountainous regions. In 1869 Emile-Oscar Ganguillet and Rudolph Kutter developed a more generally applicable discharge equation following their studies of flow in Swiss mountain streams. Toward the end of the century, systematic studies of the discharge of streams had become common. In the United States the Geological Survey, following its establishment in 1879, became the principal agency for collecting and publishing data on discharge, and by 1906 stream gauging had become nationwide.

Foundations of Oceanography

In 1807 Thomas Jefferson ordered the establishment of the U.S. Coast Survey (later Coast and Geodetic Survey and now the National Ocean Survey). Modeled after British and French agencies that had grown up in the 1700s, the agency was charged with the responsibilities of hydrographic and geodetic surveying, studies of tides, and preparation of charts. Beginning in 1842, the U.S. Navy undertook expansive oceanographic operations through its office of charts and instruments.

Lieut. Matthew Fontaine Maury promoted international cooperation in gathering meteorologic and hydrologic data at sea. In 1847 Maury compiled the first wind and current charts for the North Atlantic and in 1854 issued the first depth map to 4,000 fathoms (7,300 metres). His Physical Geography of the Sea is generally considered the first oceanographic textbook.

The voyage of the *Beagle* is remembered for Darwin's biological and geologic contributions. From his observations in the South Pacific, Darwin formulated a theory for the origin of coral reefs, which with minor changes has stood the test of time. He viewed the fringing reefs, barrier reefs, and atolls as successive stages in a developmental sequence. The volcanic islands around which the reef-building organisms are attached slowly sink, but at the same time the shallow-water organisms that form the reefs build their colonies upward so as to remain in the sunlit layers of water. With submergence of the island, what began as a fringing reef girdling a landmass at last becomes an atollenclosing a lagoon.

Laying telegraphic cables across the Atlantic called for investigations of the configuration of the ocean floor, of the currents that sweep the bottom, and of the benthonic animals that might damage the cables. The explorations of the British ships *Lightning* and *Porcupine* in 1868 and 1869 turned up surprising oceanographic information. Following closely upon these voyages, the *Challenger* was authorized to determine "the conditions of the Deep Sea throughout the Great Ocean Basins."

The *Challenger* left port in December of 1872 and returned in May 1876, after logging 127,600 kilometres (68,890 nautical miles). Under the direction of Wyville Thomson, Scottish professor of natural history, it occupied 350 stations scattered over all oceans except the Arctic. The work involved in analyzing the information gathered during the expedition was completed by Thomson's shipmate Sir John Murray, and the results filled 50 large volumes. Hundreds of new species of marine organisms were described, including new forms of life from deep waters. The temperature of water at the bottom of the oceans was found to be nearly constant below the 2,000-fathom level, averaging about 2.5 °C (36.5 °F) in the North Atlantic and 2 °C (35 °F) in the North Pacific. Soundings showed wide variations in depths of water, and from the dredgings of the bottom came new types of sediment—red clay as well as oozes made predominantly of the minute skeletons of foraminifera, radiolarians, or diatoms. Improved charts of the principal surface currents were produced, and the precise location of many oceanic islands was determined for the first time. Seventy-seven samples of seawater were taken at different stations from depths ranging downward to about 1.5 kilometres. The German-born chemist Wilhelm Dittmar conducted quantitative determinations of the seven major constituents (other than the hydrogen and oxygen of the water itself)—namely, sodium, calcium, magnesium, potassium, chloride, bromide, and sulfate. Surprisingly, the percentages of these components turned out to be nearly the same in all samples.

Efforts to analyze the rise and fall of the tides in mathematical terms reflecting the relative and constantly changing positions of Earth, Moon, and Sun, and thus to predict the tides at particular localities, has never been entirely successful because of local variations in configuration of shore and seafloor. Nevertheless, harmonic tidal analysis gives essential first approximations that are essential to tidal prediction. In 1884 a mechanical analog tidal prediction device was invented by William Ferrelof the U.S. Coast and Geodetic Survey, and improved models were used until 1965, when the work of the analog machines was taken over by electronic computers.

Atmospheric Sciences

Composition of the Atmosphere

Studies of barometric pressure by the British chemist and physicist John Dalton led him to conclude that evaporation and condensation of vapour do not involve chemical transformations. The introduction of vapour into the air by evaporation must change the average specific gravity of the air column and, without altering the height of that column, will change the reading of the barometer. In 1857 Rudolf Clausius, a German physicist, clarified the mechanics of evaporation in his kinetic theory of gases. Evaporation occurs when more molecules of a liquid are leaving its surface than returning to it, and the higher the temperature the more of these escaped molecules will be in space at any one time.

Following the invention of the hot-air balloon by the Montgolfier brothers in 1783, balloonists produced some useful information on the composition and movements of the atmosphere. In 1804 the celebrated French chemist Joseph-Louis Gay-Lussac ascended to about 7,000 metres, took samples of air, and later determined that the rarefied air at that altitude contained the same percentage of oxygen (21.49 percent) as the air on the ground. Austrian meteorologist Julius von Hann, working with data from balloon ascents and climbing in the Alps and Himalayas, concluded in 1874 that about 90 percent of all the water vapour in the atmosphere is concentrated below 6,000 metres—from which it follows that high mountains can be barriers against the transport of water vapour.

Understanding of Clouds, Fog and Dew

Most of the names given to clouds (cirrus, cumulus, stratus, nimbus, and their combinations) were coined in 1803 by the English meteorologist Luke Howard. Howard's effort was not simply taxonomic; he recognized that clouds reflect in their shapes and changing forms "the general causes which effect all the variations of the atmosphere."

After Guericke's experiments it was widely believed that water vapour condenses into cloud as soon as the air containing it cools to the dew point. That this is not necessarily so was proved by Paul-Jean Coulier of France from experiments reported in 1875. Coulier found that the sudden expansion of air in glass flasks failed to produce an artificial cloud if the air in the system was filtered through cotton wool. He concluded that dust in the air was essential to the formation of cloud in the flask.

From about the mid-1820s, efforts were made to classify precipitation in terms of the causes behind the lowering of temperature. In 1841 the American astronomer-meteorologist Elias Loomis recognized the following causes: warm air coming into contact with cold earth or water, responsible for fog; mixing of warm and cold currents, which commonly results in light rains; and sudden transport of air into high regions, as by flow up a mountain slope or by warm currents riding over an opposing current of cold air, which may produce heavy rains.

Observation and Study of Storms

Storms, particularly tropical revolving storms, were subjects of much interest. As early as 1697 some of the more spectacular features of revolving storms were recorded in William Dampier's New Voyage Round the World. On July 4, 1687, Dampier's ship survived the passage of what he called a "tuffoon" off the coast of China. The captain's vivid account of this experience clearly describes the calm central eye of the storm and the passage of winds from opposite directions as the storm moved past. In 1828 Heinrich Wilhelm Dove, a Prussian meteorologist, recognized that tropical revolving storms are traveling systems with strong winds moving counterclockwise in the Northern Hemisphere and clockwise in the Southern Hemisphere. The whirlwind character of these storms was independently established by the American meteorologist William C. Redfield in the case of the September hurricane that struck New England in 1821. He noted that in central Connecticut the trees had been toppled toward the northwest, whereas some 80 kilometres westward they had fallen in the opposite direction. Redfield identified the belt between the Equator and the tropics as the region in which hurricanes are generated, and he recognized how the tracks of these storms tend to veer eastward when they enter the belt of westerly winds at about latitude 30° N. In 1849 Sir William Reid, a British meteorologist and military engineer, studied the revolving storms that occur south of the Equator in the Indian Ocean and confirmed that they have reversed rotations and curvatures of path compared with those of the Northern Hemisphere. Capt. Henry Piddington subsequently investigated revolving storms affecting the Bay of Bengal and Arabian Sea, and in 1855 he named these cyclones in his *Sailor's Horn-book for the Laws of Storms in all Parts of the World.*

Beginning in 1835, James Pollard Espy, an American meteorologist, began extensive studies of storms from which he developed a theory to explain their sources of energy. Radially convergent winds, he believed, cause the air to rise in their area of collision. Upward movement of moist air is attended by condensation and precipitation. Latent heat released through the change of vapour to cloud or water causes further expansion and rising of the air. The higher the moist air rises the more the equilibriumof the system is disturbed, and this equilibrium cannot be restored until moist air at the surface ceases to flow toward the ascending column.

That radially convergent winds are not necessary to the rising of large air masses was demonstrated by Loomis in the case of a great storm that passed across the northeastern United States in December 1836. From his studies of wind patterns, changes of temperature, and changes in barometric pressure, he concluded that a cold northwest wind had displaced a wind blowing from the southeast by flowing under it. The southeast wind made its escape by ascending from the Earth'ssurface. Loomis had recognized what today would be called a frontal surface.

Weather and Climate

Modern meteorology began when the daily weather map was developed as a device for analysis and forecasting, and the instrument that made this kind of map possible was the electromagnetic telegraph. In the United States the first telegraph line was strung in 1844 between Washington, D.C., and Baltimore. Concurrently with the expansion of telegraphic networks, the physicist Joseph Henryarranged for telegraph companies to have meteorological instruments in exchange for current data on weather telegraphed to the Smithsonian Institution. Some 500 stations had joined this cooperative effort by 1860. The Civil War temporarily prevented further expansion, but, meanwhile, a disaster of a different order had accelerated development of synoptic meteorology in Europe. On Nov. 14, 1854, an unexpected storm wrecked British and French warships off Balaklava on the Crimean Peninsula. Had word of the approaching storm been telegraphed to this port in the Black Sea, the ships might have been saved. This mischance led in 1856 to the establishment of a national storm-warning service in France. In 1863 the Paris Observatory began publishing the first weather maps in modern format.

The first national weather service in the United States began operations in 1871 as an agency of the Department of War. The initial objective was to provide storm warnings for the Gulf and Atlantic coasts and the Great Lakes. In 1877 forecasts of temperature changes and precipitation averaged 74 percent in accuracy, as compared with 79 percent for cold-wave warnings. After 1878 daily weather maps were published.

Synoptic meteorology made possible the tracking of storm systems over wide areas. In 1868 the British meteorologist Alexander Buchan published a map showing the travels of a cyclonic depression across North America, the Atlantic, and into northern Europe. In the judgment of Sir Napier Shaw, Buchan's study marks the entry of modern meteorology, with "the weather map as its main feature and forecasting its avowed object."

In addition to weather maps, a variety of other kinds of maps showing regional variations in the components of weather and climate were produced. In 1817 Alexander von Humboldt published a map showing the distribution of mean annual temperatures over the greater part of the Northern Hemisphere. Humboldt was the first to use isothermal lines in mapping temperature. Buchan drew the first maps of mean monthly and annual pressure for the entire world. Published in 1869, these maps added much to knowledge of the general circulation of the atmosphere. In 1886 Léon-Philippe Teisserenc de Bort of France published maps showing mean annual cloudiness over the Earth for each month and the year. The first world map of precipitation showing mean annual precipitation by isohyets was the work of Loomis in 1882. This work was further refined in 1899 by the maps of the British cartographer Andrew John Herbertson, which showed precipitation for each month of the year.

Although the 19th century was still in the age of meteorologic and climatological exploration, broad syntheses of old information thus kept pace with acquisition of the new fairly well. For example, Julius Hann's massive *Handbuch der Klimatologie* ("Handbook of Climatology"), first issued in 1883, is mainly a compendium of works published in the *Meteorologische Zeitschrift*. The *Handbuch* was kept current in revised editions until 1911, and this work is still sometimes called the most skillfully written account of world climate.

The 20th Century: Modern Trends and Developments

Geologic Sciences

The development of the geologic sciences in the 20th century has been influenced by two major "revolutions." The first involves dramatic technological advances that have resulted in vastly improved instrumentation, the prime examples being the many types of highly sophisticated computerized devices. The second is centred on the development of the plate tectonics theory, which is the most profound and influential conceptual advance the Earth sciences have ever known.

Modern technological developments have affected all the different geologic disciplines. Their impact has been particularly notable in such activities as radiometric dating, experimental petrology, crystallography, chemical analysis of rocks and minerals, micropaleontology, and seismological exploration of the Earth's deep interior.

Radiometric Dating

In 1905, shortly after the discovery of radioactivity, the American chemist Bertram Boltwood suggested that lead is one of the disintegration products of uranium, in which case the older a uranium-bearing mineral the greater should be its proportional part of lead. Analyzing specimens whose relative geologic ages were known, Boltwood found that the ratio of lead to uranium did indeed increase with age. After estimating the rate of this radioactive change, he calculated that the absolute ages of his specimens ranged from 410 million to 2.2 billion years. Though his figures were too high by about 20 percent, their order of magnitude was enough to dispose of the short scale of geologic time proposed by Lord Kelvin.

Versions of the modern mass spectrometer were invented in the early 1920s and 1930s, and during World War II the device was improved substantially to help in the development of the atomic bomb. Soon after the war, Harold C. Urey and G.J. Wasserburg applied the mass spectrometer to the study of geochronology. This device separates the different isotopes of the same element and can measure the variations in these isotopic abundances to within one part in 10,000. By determining the amount of the parent and daughter isotopes present in a sample and by knowing their rate of radioactive decay (each radioisotope has its own decay constant), the isotopic age of the sample can be calculated. For dating minerals and rocks, investigators commonly use the following couplets of parent and daughter isotopes: thorium-232–lead-208, uranium-235–lead-207, samarium-147–neodymium-143, rubidium-87–strontium-87, potassium-40–argon-40, and argon-40–argon-39. The SHRIMP (Sensitive High Resolution Ion Microprobe) enables the accurate determination of the uranium-lead age of the mineral zircon, and this has revolutionized the understanding of the isotopic age of formation of zircon-bearing igneous granitic rocks. Another technological development is the ICP-MS (Inductively Coupled Plasma Mass Spectrometer), which is able to provide the isotopic age of the minerals zircon, titanite, rutile, and monazite. These minerals are common to many igneous and metamorphic rocks.

Such techniques have had an enormous impact on scientific knowledge of Earth history because precise dates can now be obtained on rocks in all orogenic (mountain) belts ranging in age from the early Archean (about 4 billion years old) to the early Neogene (roughly 20 million years old). The oldest known rocks on Earth, estimated at 4.28 billion years old, are the faux amphibolite volcanic deposits of the Nuvvuagittuq greenstone belt in Quebec, Canada. A radiometric dating technique that measures the ratio of the rare earth elements neodymium and samarium present in a rock sample was used to produce the estimate. Also, by extrapolating backward in time to a situation when there was no lead that had been produced by radiogenic processes, a figure of about 4.6 billion years is obtained for the minimum age of the Earth.

Experimental Study of Rocks

Experimental petrology began with the work of Jacobus Henricus van 't Hoff, one of the founders of physical chemistry. Between 1896 and 1908 he elucidated the complex sequence of chemical reactions attending the precipitation of salts (evaporites) from the evaporation of seawater. Van 't Hoff's aim was to explain the succession of mineral salts present in Permian rocks of Germany. His success at producing from aqueous solutions artificial minerals and rocks like those found in natural salt deposits stimulated studies of minerals crystallizing from silicate melts simulating the magmas from which igneous rocks have formed. Working at the Geophysical Laboratory of the Carnegie Institution of Washington, D.C., Norman L. Bowen conducted extensive phase-equilibrium studies of silicate systems, brought together in his Evolution of the Igneous Rocks.

Experimental petrology, both at the low-temperature range explored by van 't Hoff and in the high ranges of temperature investigated by Bowen, continues to provide laboratory ev-idence for interpreting the chemical history of sedimentary and igneous rocks. Experimental petrology also provides valuable data on the stability limits of individual metamorphic minerals and of the reactions between different minerals in a wide variety of chemical systems. These experiments are carried out at elevated temperatures and pressures that simulate those operating in different levels of the Earth's crust. Thus the metamorphic petrologist today can compare the minerals and mineral assemblages found in natural rocks with comparable examples produced in the laboratory, the pressure–temperature limits of which have been well defined by experimental petrology.

Another branch of experimental science relates to the deformation of rocks. In 1906 the American physicist P.W. Bridgman developed a technique for subjecting rock samples to high pressures similar to those deep in the Earth. Studies of the behaviour of rocks in the laboratory have shown that their strength increases with confining pressure but decreases with rise in temperature. Down to depths of a few kilometres the strength of rocks would be expected to increase. At greater depths the temperature effect should become dominant, and response to stress should result in flow rather than fracture of rocks. In 1959 two American geologists, Marion King Hubbert and William W. Rubey, demonstrated that fluids in the pores of rock may reduce internal friction and permit gliding over nearly horizontal planes of the large overthrust blocks associated with folded mountains. More recently the Norwegian petrologist Hans Ramberg performed many experiments with a large centrifuge that produced a negative gravity effect and thus was able to create structures simulating salt domes, which rise because of the relatively low density of the salt in comparison with that of surrounding rocks. With all these deformation experiments, it is necessary to scale down as precisely as possible variables such as the time and velocity of the experiment and the viscosity and temperature of the material from the natural to the laboratory conditions.

Crystallography

In the 19th century crystallographers were able to study only the external form of minerals, and it was not until 1895 when the German physicist Wilhelm Conrad Röntgen discovered X-rays that it became possible to consider their internal structure. In 1912 another German physicist, Max von Laue, realized that X-rays were scattered and deflected at regular angles when they passed through a copper sulfate crystal, and so he produced the first X-ray diffraction pattern on a photographic film. A year later William Bragg of Britain and his son Lawrence perceived that such a pattern reflects the layers of atoms in the crystal structure, and they succeeded in determining for the first time the atomic crystal structure of the mineral halite (sodium chloride). These discoveries had a long-lasting influence on crystallography because they led to the development of the X-ray powder diffractometer, which is now widely used to identify minerals and to ascertain their crystal structure.

The Chemical Analysis of Rocks and Minerals

Advanced analytic chemical equipment has revolutionized the understanding of the composition of rocks and minerals. For example, the XRF (X-Ray Fluorescence) spectrometer can quantify the major and trace element abundances of many chemical elements in a rock sample down to parts-per-million concentrations. This geochemical method has been used to differentiate successive stages of igneous rocks in the plate-tectonic cycle. The metamorphic petrologist can use the bulk composition of a recrystallized rock to define the structure of the original rock, assuming that no structural change has occurred during the metamorphic process. Next, the electron microprobe bombards a thin microscopic slice of a mineral in a sample with a beam of electrons, which can determine the chemical composition of the mineral almost instantly. This method has wide applications in, for example, the fields of industrial mineralogy, materials science, igneous geochemistry, and metamorphic petrology.

Micropaleontology

Microscopic fossils, such as ostracods, foraminifera, and pollen grains, are common in sediments of the Mesozoic and Cenozoic eras (from about 251 million years ago to the present). Because the rock chips brought up in oil wells are so small, a high-resolution instrument known as a scanning electron microscope had to be developed to study the microfossils. The classification of microfossils of organisms that lived within relatively short time spans has enabled Mesozoic-Cenozoic sediments to be subdivided in remarkable detail. This technique also has had a major impact on the study of Precambrian life (i.e., organisms that existed more than 542 million years ago). Carbonaceous spheroids and filaments about 7–10 millimetres (0.3–0.4 inch) long are recorded in 3.5 billion-year-old sediments in the Pilbara region of northwestern Western Australia and in the lower Onverwacht Series of the Barberton belt in South Africa; these are the oldest reliable records of life on Earth.

Seismology and the Structure of the Earth

Earthquake study was institutionalized in 1880 with the formation of the Seismological Society of Japan under the leadership of the English geologist John Milne. Milne and his associates invented the first accurate seismographs, including the instrument later known as the Milne seismograph. Seismology has revealed much about the structure of the Earth's core, mantle, and crust. The English seismologist Richard Dixon Oldham's studies of earthquake records in 1906 led to the discovery of the Earth's core. From studies of the Croatian quake of Oct. 8, 1909, the geophysicist Andrija Mohorovičić discovered the discontinuity (often called the Moho) that separates the crust from the underlying mantle.

Today there are more than 1,000 seismograph stations around the world, and their data are used to compile seismicity maps. These maps show that earthquake epicentres are aligned in narrow, continuous belts along the boundaries of lithospheric plates. The earthquake foci outline the mid-oceanic ridges in the Atlantic, Pacific, and Indian oceans where the plates separate, while around the margins of the Pacific where the plates converge, they lie in a dipping plane, or Benioff zone, that defines the position of the subducting plate boundary to depths of about 700 kilometres.

Since 1950, additional information on the crust has been obtained from the analysis of artificial tremors produced by chemical explosions. These studies have shown that the Moho is present under all continents at an average depth of 35 kilometres and that the crust above it thickens under young mountain ranges to depths of 70 kilometres in the Andes and the Himalayas. In such investigations the reflections of the seismic waves generated from a series of "shot" points are also recorded, and this makes it possible to construct a profile of the subsurface structure. This is seismic reflection profiling, the main method of exploration used by the petroleum industry. During the late 1970s a new technique for generating seismic waves was invented: thumping and vibrating the surface of the ground with a gas-propelled piston from a large truck.

The Theory of Plate Tectonics

Plate tectonics has revolutionized virtually every discipline of the Earth sciences since the late 1960s and early 1970s. It has served as a unifying model or paradigm for explaining geologic phenomena that were formerly considered in unrelated fashion. Plate tectonics describes seismic activity, volcanism, mountain building, and various other Earth processes in terms of the structure and mechanical behaviour of a small number of enormous rigid plates thought to constitute the outer part of the planet (i.e., the lithosphere). This all-encompassing theory grew out of observations and ideas about continental drift and seafloor spreading.

In 1912 the German meteorologist Alfred Wegener proposed that throughout most of geologic time there was only one continental mass, which he named Pangea. At some time during the Mesozoic Era, Pangaea fragmented and the parts began to drift apart. Westward drift of the Americas opened the Atlantic Ocean, and the Indian block drifted across the Equator to join with Asia. In 1937 the South African Alexander Du Toit modified Wegener's hypothesis by suggesting the existence of two primordial continents: Laurasia in the north and Gondwanaland in the south. Aside from the congruency of continental shelf margins across the Atlantic, proponents of continental drift have amassed impressive geologic evidence to support their views. Similarities in fossil terrestrial organisms in pre-Cretaceous (older than about 146 million years) strata of Africa and South America and in pre-Jurassic rocks (older than about 200 million years) of Australia, India, Madagascar, and Africa are explained if these continents were formerly connected but difficult to account for otherwise. Fitting the Americas with the continents across the Atlantic brings together similar kinds of rocks and structures. Evidence of widespread glaciation during the late Paleozoic is found in Antarctica, southern South America, southern Africa, India, and Australia. If these continents were formerly united around the South Polar region, this glaciation becomes explicable as a unified sequence of events in time and space.

Interest in continental drift heightened during the 1950s as knowledge of the Earth's magnetic field during the geologic past developed from the studies of Stanley K. Runcorn, Patrick M.S. Blackett, and others. Ferromagnetic minerals such as magnetite acquire a permanent magnetization when they crystallize as components of igneous rock. The direction of their magnetization is the same as the direction of the Earth's magnetic field at the place and time of crystallization. Particles of magnetized minerals released from their parent igneous rocks by weathering may later realign themselves with the existing magnetic field at the time these particles are incorporated into sedimentary deposits. Studies of the remanent magnetism in suitable rocks of different ages from over the world indicate that the magnetic poles were in different places at different times. The polar wandering curves are different for the several continents, but in important instances these differences are reconciled on the assumption that continents now separated were formerly joined. The curves for Europe and North America, for example, are reconciled by the assumption that America has drifted about 30° westward relative to Europe since the Triassic Period (approximately 251 million to 200 million years ago).

In the early 1960s a major breakthrough in understanding the way the modern Earth works came from two studies of the ocean floor. First, the American geophysicists Harry H. Hess and Robert S. Dietz suggested that new ocean crust was formed along mid-oceanic ridges between separating continents; and second, Drummond H. Matthews and Frederick J. Vine of Britain proposed that the new oceanic crust acted like a magnetic tape recorder insofar as magnetic anomaly strips parallel to the ridge had been magnetized alternately in normal and reversed order, reflecting the changes in polarity of the Earth's magnetic field. This theory of seafloor spreading then needed testing, and the opportunity arose from major advances in deepwater drilling technology. The Joint Oceanographic Institutions Deep Earth Sampling (JOIDES) project began in 1969, continued with the Deep Sea Drilling Project (DSDP), and, since 1976, with the International Phase of Ocean Drilling (IPOD) project. These projects have produced more than 500 boreholes in the floor of the world's oceans, and the results have been as outstanding as the plate-tectonic theory itself. They confirm that the oceanic crust is everywhere younger than about 200 million years and that the stratigraphic age determined by micropaleontology of the overlying oceanic sediments is close to the age of the oceanic crust calculated from the magnetic anomalies.

The plate-tectonic theory, which embraces both continental drift and seafloor spreading, was formulated in the mid-1960s by the Canadian geologist J. Tuzo Wilson, who described the network of mid-oceanic ridges, transform faults, and subduction zones as boundaries separating an evolving mosaic of enormous plates, and who proposed the idea of the opening and closing of oceans and eventual production of an orogenic belt by the collision of two continents.

Up to this point, no one had considered in any detail the implications of the plate-tectonic theory for the evolution of continental orogenic belts; most thought had been devoted to the oceans. In 1969 John Dewey of the University of Cambridge outlined an analysis of the Caledonian-Appalachian orogenic belts in terms of a complete plate-tectonic cycle of events, and this provided a model for the interpretation of other pre-Mesozoic (Paleozoic and Precambrian) belts. Even the oldest volcano-sedimentary rocks on Earth, in the 3.8 billion-year-old Isua belt in West Greenland, have been shown by geologists from the Tokyo Institute of Technology to have formed in a plate-tectonic setting—i.e., in a trench or mouth of a subduction zone.

Hydrologic Sciences

Water Resources and Seawater Chemistry

Quantitative studies of the distribution of water have revealed that an astonishingly small part of the Earth's water is contained in lakes and rivers. Ninety-seven percent of all the water is in the oceans, and, of the fresh water constituting the remainder, three-fourths is locked up in glacial ice and most of the rest is in the ground. Approximate figures are also now available for the amounts of water involved in the different stages of the hydrologic cycle. Of the 859 millimetres of annual global precipitation, 23 percent falls on the lands; but only about a third of the precipitation on the lands runs directly back to the sea, the remainder being recycled through the atmosphere by evaporation and transpiration. Subsurface groundwater accumulates by infiltration of rainwater into soil and bedrock. Some may run off into rivers and lakes, and some may reemerge as springs or aquifers. Advanced techniques are used extensively in groundwater studies nowadays. The rate of groundwater flow, for example, can be calculated from the breakdown of radioactive carbon-14 by measuring the time it takes for rainwater to pass through the ground, while numerical modeling is used to study heat and mass transfer in groundwater.

High-precision equipment is used for measuring down-hole temperature, pressure, flow rate, and water level. Groundwater hydrology is important in studies of fractured reservoirs, subsidence resulting from fluid withdrawal, geothermal resource exploration, radioactive waste disposal, and aquifer thermal-energy storage.

Chemical analyses of trace elements and isotopes of seawater are conducted as part of the Geochemical Ocean Sections (Geosecs) program. Of the 92 naturally occurring elements, nearly 80 have been detected in seawater or in the organisms that inhabit it, and it is thought to be only a matter of time until traces of the others are detected. Contrary to the idea widely circulated in the older literature of oceanography, that the relative proportions of the oceans' dissolved constituents are constant, investigations since 1962 have revealed statistically significant variations in the ratios of calcium and strontium to chlorinity. The role of organisms as influences on the composition of seawater has become better understood with advances in marine biology. It is now known that plants and animals may collect certain elements to concentrations as much as 100,000 times their normal amounts in seawater. Abnormally high concentrations of beryllium, scandium, chromium, and iodine have been found in algae; of copper and arsenic in both the soft and skeletal parts of invertebrate animals; and of zirconium and cerium in plankton.

Desalinization, Tidal Power and Minerals from the Sea

For ages a source of food and common salt, the sea is increasingly becoming a source of water, chemicals, and energy. In 1967 Key West, Fla., became the first U.S. city to be supplied solely by water from the sea, drawing its supplies from a plant that produces more than 2 million gallons of refined water daily. Magnesia was extracted from the Mediterranean in the late 19th century; at present nearly all the magnesium metal used in the United States is mined from the sea at Freeport, Texas. Many ambitious schemes for using tidal power have been devised, but the first major hydrographic project of this kind was not completed until 1967, when a dam and electrical generating equipment were installed across the Rance River in Brittany. The seafloor and the strata below the continental shelves are also sources of mineral wealth. Concretions of manganese oxide, evidently formed in the process of subaqueous weathering of volcanic rocks, have been found in dense concentrations with a total abundance of 1011 tons. In addition to the manganese, these concretions contain copper, nickel, cobalt, zinc, and molybdenum. To date, oil and gas have been the most valuable products to be produced from beneath the sea.

Ocean Bathymetry

Modern bathymetric charts show that about 20 percent of the surfaces of the continents are submerged to form continental shelves. Altogether the shelves form an area about the size of Africa. Continental slopes, which slant down from the outer edges of the shelves to the abyssal plains of the seafloor, are nearly everywhere furrowed by submarine canyons. The depths to which these canyons have been cut below sea level seem to rule out the possibility that they are drowned valleys cut by ordinary streams. More likely, the canyons were eroded by turbidity currents, dense mixtures of mud and water that originate as mudslides in the heads of the canyons and pour down their bottoms.

Profiling of the Pacific basin prior to and during World War II resulted in the discovery of hundreds of isolated eminences rising 1,000 or more metres above the floor. Of particular interest were seamounts in the shape of truncated cones, whose flat tops rise to between 1.6 kilometres and a few hundred metres below the surface. Harry H. Hess interpreted the flat-topped seamounts (guyots) as volcanic mountains planed off by action of waves before they subsided to their present depths. Subsequent drilling in guyots west of Hawaii confirmed this view; samples of rocks from the tops contained fossils of Cretaceous age representing reef-building organisms of the kind that inhabit shallow water.

Ocean Circulation, Currents and Waves

Early in the 20th century Vilhelm Bjerknes, a Norwegian meteorologist, and V. Walfrid Ekman, a Swedish physical oceanographer, investigated the dynamics of ocean circulation and developed theoretical principles that influenced subsequent studies of currents in the sea. Bjerknes showed that very small forces resulting from pressure differences caused by nonuniform density of seawater can initiate and maintain fluid motion. Ekman analyzed the influence of winds and the Earth's rotation on currents. He theorized that in a homogeneous medium the frictional effects of winds blowing across the surface would cause movement of successively lower layers of water, the deeper the currents so produced the less their velocity and the greater their deflection by the Coriolis effect (an apparent force due to the Earth's rotation that causes deflection of a moving body to the right in the Northern Hemisphere and to the left in the Southern Hemisphere), until at some critical depth an induced current would move in a direction opposite to that of the wind.

Results of many investigations suggest that the forces that drive the ocean currents originate at the interface between water and air. The direct transfer of momentum from the atmosphere to the sea is doubtless the most important driving force for currents in the upper parts of the ocean. Next in importance are differential heating, evaporation, and precipitation across the air-sea boundary, altering the density of seawater and thus initiating movement of water masses with different densities. Studies of the properties and motion of water at depth have shown that strong currents also exist in the deep sea and that distinct types of water travel far from their geographic sources. For example, the highly saline water of the Mediterranean that flows through the Strait of Gibraltar has been traced over a large part of the Atlantic, where it forms a deepwater stratum that is circulated far beyond that ocean in currents around Antarctica.

Improvements in devices for determining the motion of seawater in three dimensions have led to the discovery of new currents and to the disclosure of unexpected complexities in the circulation of the oceans generally. In 1951 a huge countercurrent moving eastward across the Pacific was found below depths as shallow as 20 metres, and in the following year an analogous equatorial undercurrent was discovered in the Atlantic. In 1957 a deep countercurrent was detected beneath the Gulf Stream with the aid of subsurface floats emitting acoustic signals.

Since the 1970s Earth-orbiting satellites have yielded much information on the temperature distribution and thermal energy of ocean currents such as the Gulf Stream. Chemical analyses from Geosecs makes possible the determination of circulation paths, speeds, and mixing rates of ocean currents.

Surface waves of the ocean are also exceedingly complex, at most places and times reflecting the coexistence and interferences of several independent wave systems. During World War II, interest in forecasting wave characteristics was stimulated by the need for this critical information in the planning of amphibious operations. The oceanographers H.U. Sverdrup and Walter Heinrich Munkcombined theory and empirical relationships in developing a method of forecasting "significant wave height"—the average height of the highest third of the waves in a wave train. Subsequently this method was improved to permit wave forecasters to predict optimal routes for mariners. Forecasting of the most destructive of all waves, tsunamis, or "tidal waves," caused by submarine quakes and volcanic eruptions, is another recent development. Soon after 159 persons were killed in Hawaii by the tsunami of 1946, the U.S. Coast and Geodetic Survey established a seismic sea-wave warning system. Using a seismic network to locate epicentres of submarine quakes, the installation predicts the arrival of tsunamis at points around the Pacific basin often hours before the arrival of the waves.

Glacier Motion and the High-latitude Ice Sheets

Beginning around 1948, principles and techniques in metallurgy and solid-state physics were brought to bear on the mechanics of glacial movements. Laboratory studies showed that glacial ice deforms like other crystalline solids (such as metals) at temperatures near the melting point. Continued stress produces permanent deformation. In addition to plastic deformation within a moving glacier, the glacier itself may slide over its bed by mechanisms involving pressure melting and refreezing and accelerated plastic flow around obstacles. The causes underlying changes in rate of glacial movement, in particular spectacular accelerations called surges, require further study. Surges involve massive transfer of ice from the upper to the lower parts of glaciers at rates of as much as 20 metres a day, in comparison with normal advances of a few metres a year.

As a result of numerous scientific expeditions into Greenland and Antarctica, the dimensions of the remaining great ice sheets are fairly well known from gravimetric and seismic surveys. In parts of both continents it has been determined that the base of the ice is below sea level, probably due at least in part to subsidence of the crust under the weight of the caps. In 1966 a borehole was drilled 1,390 metres to bedrock on the North Greenlandice sheet, and two years later a similar boring of 2,162 metres was cut through the Antarctic ice at Byrd Station. From the study of annual incremental layers and analyses of oxygen isotopes, the bottom layers of ice cored in Greenland were estimated to be more than 150,000 years old, compared with 100,000 years for the Antarctic core. With the advent of geochemical dating of rocks it has become evident that the Ice Age, which in the earlier part of the century was considered to have transpired during the Quaternary Period, actually began much earlier. In Antarctica, for example, potassium-argon age determinations of lava overlying glaciated surfaces and sedimentary deposits of glacial origin show that glaciers existed on this continent at least 10 million years ago.

The study of ice sheets has benefited much from data produced by advanced instruments, computers, and orbiting satellites. The shape of ice sheets can be determined by numerical modeling, their heat budget from thermodynamic calculations, and their thickness with radar techniques. Colour images from satellites show the temperature distribution across the polar regions, which can be compared with the distribution of land and sea ice.

Atmospheric Sciences

Probes, Satellites and Data Transmission

Kites equipped with meteorgraphs were used as atmospheric probes in the late 1890s, and in 1907 the U.S. Weather Bureau recorded the ascent of a kite to 7,044 metres above Mount Weather, Virginia.

In the 1920s the radio replaced the telegraph and telephone as the principal instrument for transmitting weather data. By 1936 the radio meteorgraph (radiosonde) was developed, with capabilities of sending signals on relative humidity, temperature, and barometric pressure from unmanned balloons. Experimentation with balloons up to altitudes of about 31 kilometres showed that columns of warm air may rise more than 1.6 kilometres above the Earth's surface and that the lower atmosphere is often stratified, with winds in the different layers blowing in different directions. During the 1930s airplanes began to be used for observations of the weather, and the years since 1945 have seen the development of rockets and weather satellites. TIROS (Television Infra-Red Observation Satellite), the world's first all-weather satellite, was launched in 1960, and in 1964 the Nimbus Satellite of the United States National Aeronautics and Space Administration (NASA) was rocketed into near-polar orbit.

There are two types of weather satellites: polar and geostationary. Polar satellites, like Nimbus, orbit the Earth at low altitudes of a few hundred kilometres, and, because of their progressive drift, they produce a photographic coverage of the entire Earth every 24 hours. Geostationary satellites, first sent up in 1966, are situated over the Equator at altitudes of about 35,000 kilometres and transmit data at regular intervals. Much information can be derived from the data collected by satellites. For example, wind speed and direction are measured from cloud trajectories, while temperature and moisture profiles of the atmosphere are calculated from infrared data.

Weather Forecasting

Efforts at incorporating numerical data on weather into mathematical formulas that could then be used for forecasting were initiated early in the century at the Norwegian Geophysical Institute. Vilhelm Bjerknes and his associates at Bergen succeeded in devising equations relating the measurable components of weather, but their complexity precluded the rapid solutions needed for forecasting. Out of their efforts, however, came the polar front theory for the origin of cyclones and the now-familiar names of cold front, warm front, and stationary front for the leading edges of air masses.

In 1922 the British mathematician Lewis Fry Richardson demonstrated that the complex equations of the Norwegian school could be reduced to long series of simple arithmetic operations. With no more than the desk calculators and slide rules then available, however, the solution of a problem in procedure only raised a new one in manpower. In 1946 the mathematician John von Neumann and his fellow workers at the Institute for Advanced Study, in Princeton, N.J., began work on an electronic device to do the computation faster than the weather developed. Four years later the von Neumann group could claim that, given adequate data, their computer could forecast the weather as well as a weatherman. Present-day numerical weather forecasting is achieved with the help of advanced computer analysis.

Cloud Physics

Studies of cloud physics have shown that the nuclei around which water condenses vary widely in their degree of concentration and areal distribution, ranging from six per cubic centimetre over the oceans to more than 4 million per cubic centimetre in the polluted air of some cities. The droplets that condense on these foreign particles may be as small as 0.001 centimetre in diameter. Raindrops apparently may form directly from the coalescence of these droplets, as in the case of tropical rains, or in the temperate zones through the intermediary of ice crystals. According to the theory of Tor Bergsonand Walter Findeisen, vapour freezing on ice crystals in the clouds enlarges the crystals until they fall. What finally hits the ground depends on the temperature of air below the cloud—if below freezing, snow; if above, rain.

Properties and Structure of the Atmosphere

Less than a year after the space age began with the launching of the Soviet Sputnik I in 1957, the U.S. satellite Explorer I was sent into orbit with a Geiger counter for measuring the intensity of cosmic radiation at different levels above the ground. At altitudes around 1,000 kilometres this instrument ceased to function due to saturation by charged particles. This and subsequent investigations showed that a zone of radiation encircles the world between about latitude 75° N and 75° S, with maximum intensities at 5,000 and 16,000 kilometres. Named after the American physicist James Van Allen, a leading investigator of this portion of the Earth's magnetosphere, these zones are responsive to events taking place on the Sun. The solar wind, a stream of atomic particles emanating from the Sun in all directions, seems to be responsible for the electrons entrapped in the Van Allen region as well as for the teardrop shape of the magnetosphere as a whole, with its tail pointing always away from the Sun.

In 1898 Teisserenc de Bort, studying variations of temperature at high altitudes with the aid of balloons, discovered that at elevations of about 11 kilometres the figure for average decrease of temperature with height (about 5.5 °C per 1,000 metres of ascent) dropped and the value remained nearly constant at around −55 °C. He named the atmospheric zones below and above this temperature boundary the troposphere and the stratosphere.

Toward the end of World War II the B-29 Superfortress came into use as the first large aircraft to cruise regularly at 10,000 metres. Heading westward from bases in the Pacific, these planes sometimes encountered unexpected head winds that slowed their flight by as much as 300 kilometres per hour. The jet streams, as these high-altitude winds were named, have been found to encircle the Earth following wavy courses and moving from west to east at velocities ranging upward to 500 kilometres per hour. Aircraft have also proved useful in studies of the structure and dynamics of tropical hurricanes. Following the destruction wrought to the Atlantic Coast of the United States in 1955 by hurricanes Connie and Diane, a national centre was established in Florida with the missions of locating and tracking and, it is hoped, of learning how to predict the paths of hurricanes and to dissipate their energy.

Weather Modification

As late as the 1890s experiments were conducted in the United States in the hope of producing rainby setting off charges of dynamite lofted by balloons or kites. No positive results were reported, however. More promising were the cloud-seeding experiments of the 1940s, in which silver iodide was released into clouds as smoke or solid carbon dioxide broadcast into clouds from airplanes. The results are still uncertain for increasing precipitation. The lessons learned from cloud seeding, however, have had other successful applications, such as the dispersal of low-level supercooled fog at airports (the first system designed for this purpose, the Turboclair fog-dissipation system, was set up in 1970 at Orly airport in Paris).

The inadvertent weather modification that has followed industrialization and the building of large cities has, however, already produced measurable changes in local climate and may someday produce effects more widespread. The introduction of some 12 billion tons of carbon dioxide into the atmosphere each year from the burning of fuels may in time raise the Earth's average temperature. Cities affect the flow of wind, warm the atmosphere over them, and send pollutants into the sky. Updrafts and an abundance of condensation nuclei may increase rainfall and winter fog and reduce sunshine and daylight.

Geography

Geography is a field of science devoted to the study of the lands, features, inhabitants, and phenomena of the Earth and planets. Geography is an all-encompassing discipline that seeks an understanding of Earth and its human and natural complexities—not merely where objects are, but also how they have changed and come to be.

Geography is often defined in terms of two branches: human geography and physical geography. Human geography deals with the study of people and their communities, cultures, economies, and interactions with the environment by studying their relations with and across space and place. Physical geography deals with the study of processes and patterns in the natural environment like the atmosphere, hydrosphere, biosphere, and geosphere.

The four historical traditions in geographical research are: spatial analyses of natural and the human phenomena, area studies of places and regions, studies of human-land relationships, and the Earth sciences. Geography has been called "the world discipline" and "the bridge between the human and the physical sciences".

Physical map of Earth with political borders as of 2016.

Geography is a systematic study of the Universe and its features. Traditionally, geography has been associated with cartography and place names. Although many geographers are trained in toponymy and cartology, this is not their main preoccupation. Geographers study the space and the temporal database distribution of phenomena, processes, and features as well as the interaction of humans and their environment. Because space and place affect a variety of topics, such as economics, health, climate, plants and animals, geography is highly interdisciplinary. The interdisciplinary nature of the geographical approach depends on an attentiveness to the relationship between physical and human phenomena and its spatial patterns.

> "Names of places are not geography know by heart a whole gazetteer full of them would not, in itself, constitute anyone a geographer. Geography has higher aims than this: it seeks to classify phenomena (alike of the natural and of the political world, in so far as it treats of the latter), to compare, to generalize, to ascend from effects to causes, and, in doing so, to trace out the laws of nature and to mark their influences upon man. This is 'a description of the world'—that is Geography. In a word Geography is a Science—a thing not of mere names but of argument and reason, of cause and effect".
>
> — *William Hughes*

> "Just as all phenomena exist in time and thus have a history, they also exist in space and have a geography".
>
> — *United States National Research Council*

Geography as a discipline can be split broadly into two main subsidiary fields: human geography and physical geography. The former largely focuses on the built environment and how humans create, view, manage, and influence space. The latter examines the natural environment, and how organisms, climate, soil, water, and landforms produce and interact. The difference between these approaches led to a third field, environmental geography, which combines physical and human geography and concerns the interactions between the environment and humans.

Branches

Physical Geography

Physical geography (or physiography) focuses on geography as an Earth science. It aims to understand the physical problems and the issues of lithosphere, hydrosphere, atmosphere, pedosphere, and global flora and fauna patterns (biosphere).

Physical geography can be divided into many broad categories, including:

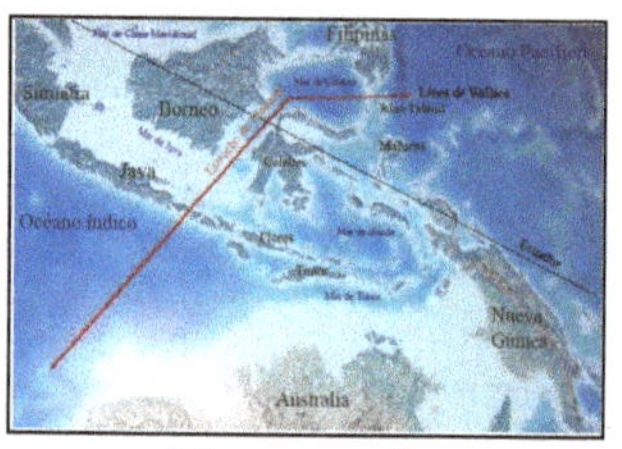

Biogeography

Climatology & meteorology

Coastal geography

Environmental management

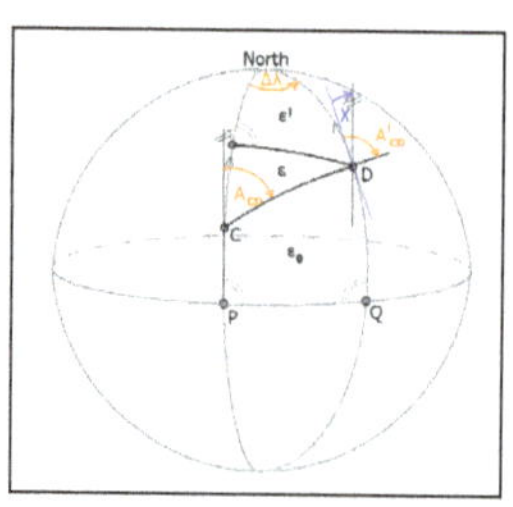

Geodesy

Geomorphology

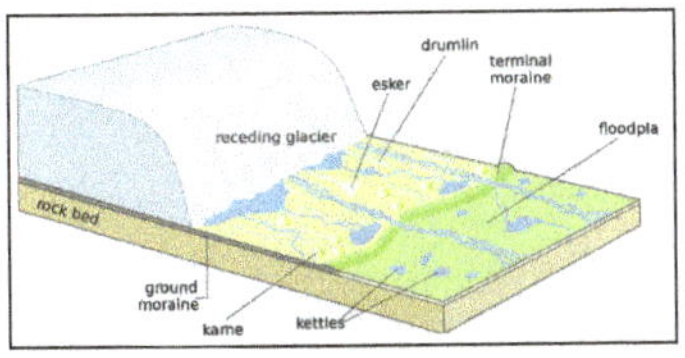

Glaciology

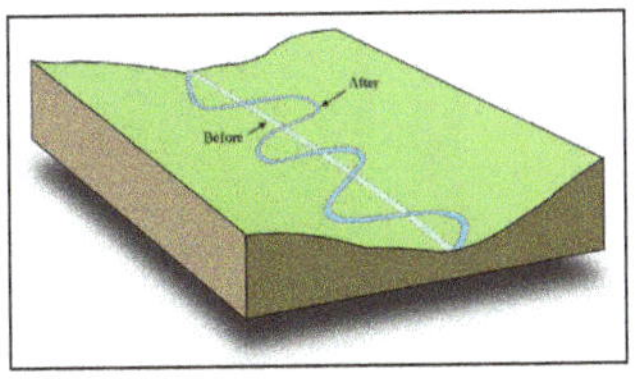
Hydrology & hydrography

Landscape ecology

Oceanography

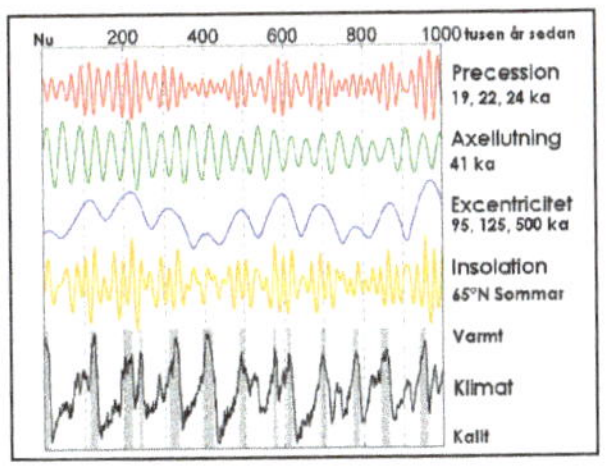

Quaternary science

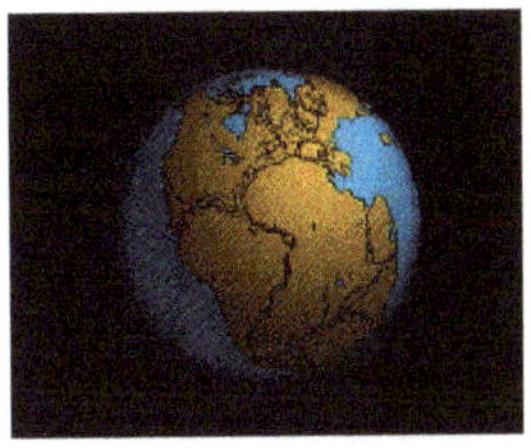
Palaeogeography

Human Geography

Human geography is a branch of geography that focuses on the study of patterns and processes that shape the human society. It encompasses the human, political, cultural, social, and economic aspects.

Human geography can be divided into many broad categories, such as:

Cultural geography

Development geography

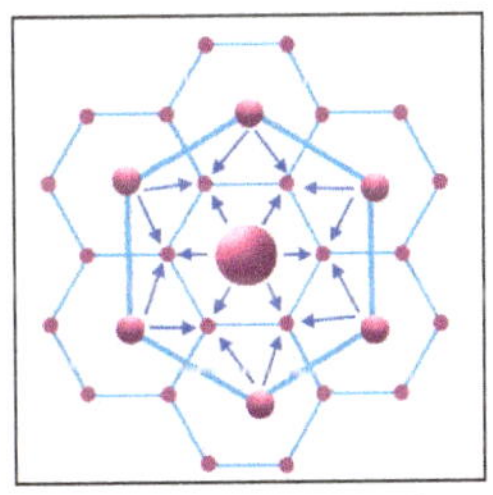
Economic geography

Health geography

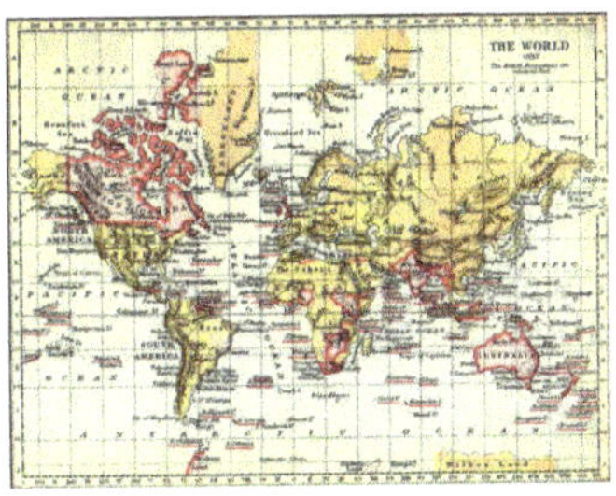
Historical & Time geog.

Political geog. & Geopolitics

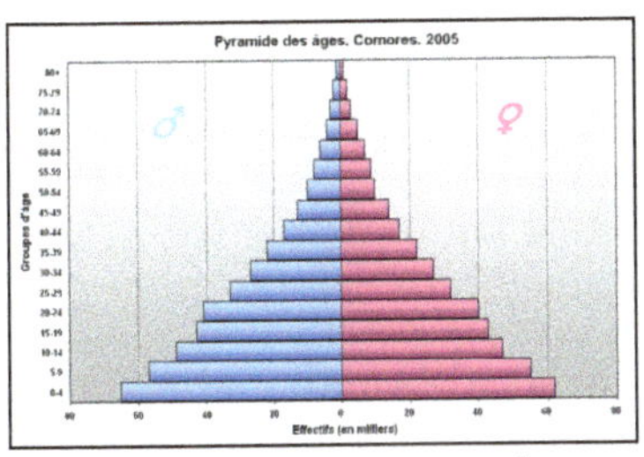

Pop. geog. or Demography

Religion geography

Social geography

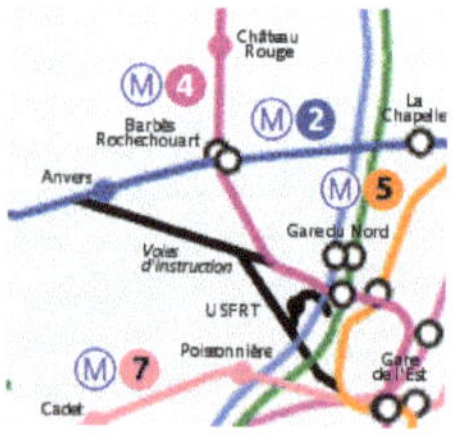

Transportation geography

Tourism geography

Urban geography

Various approaches to the study of human geography have also arisen through time and include:

- Behavioral geography
- Feminist geography
- Culture theory
- Geosophy

Integrated Geography

Integrated geography is concerned with the description of the spatial interactions between humans and the natural world. It requires an understanding of the traditional aspects of physical and human geography, as well as the ways that human societies conceptualize the environment. Integrated geography has emerged as a bridge between human and physical geography, as a result of the increasing specialisation of the two sub-fields. Furthermore, as the human relationship with the environment has changed as a result of globalization and technological change, a new approach was needed to understand the changing and dynamic relationship. Examples of areas of research in environmental geography include: emergency management, environmental management, sustainability, and political ecology.

Geomatics

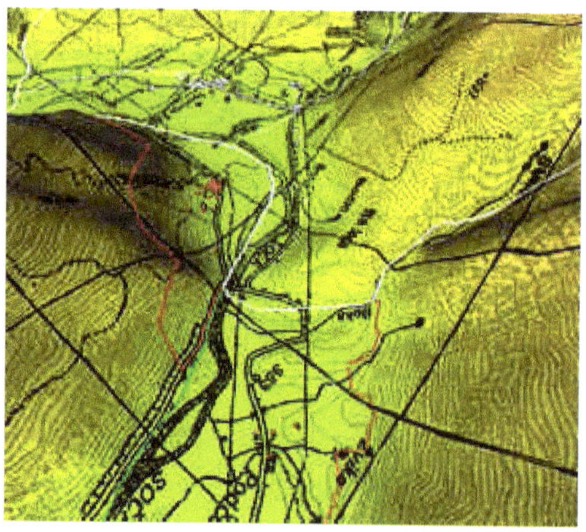
Digital Elevation Model (DEM)

Geomatics is concerned with the application of computers to the traditional spatial techniques used in cartography and topography. Geomatics emerged from the quantitative revolution in geography in the mid-1950s. Today, geomatics methods include spatial analysis, geographic information systems (GIS), remote sensing, and global positioning systems (GPS). Geomatics has led to a revitalization of some geography departments, especially in Northern America where the subject had a declining status during the 1950s.

Regional Geography

Regional geography is concerned with the description of the unique characteristics of a particular region such as its natural or human elements. The main aim is to understand, or define the uniqueness, or character of a particular region that consists of natural as well as human elements. Attention is paid also to regionalization, which covers the proper techniques of space delimitation into regions.

Related Fields

- Urban planning, regional planning, and spatial planning: Use the science of geography to assist in determining how to develop (or not develop) the land to meet particular criteria, such as safety, beauty, economic opportunities, the preservation of the built or natural heritage, and so on. The planning of towns, cities, and rural areas may be seen as applied geography.
- Regional science: In the 1950s, the regional science movement led by Walter Isard arose to provide a more quantitative and analytical base to geographical questions, in contrast to the descriptive tendencies of traditional geography programs. Regional science comprises the body of knowledge in which the spatial dimension plays a fundamental role, such as regional economics, resource management, location theory, urban and regional planning, transport and communication, human geography, population distribution, landscape ecology, and environmental quality.
- Interplanetary Sciences: While the discipline of geography is normally concerned with the Earth, the term can also be informally used to describe the study of other worlds, such as the planets of the Solar System and even beyond. The study of systems larger than the Earth itself usually forms part of Astronomy or Cosmology. The study of other planets is usually called planetary science. Alternative terms such as areology (the study of Mars) have been proposed but are not widely used.

Techniques

As spatial interrelationships are key to this synoptic science, maps are a key tool. Classical cartography has been joined by a more modern approach to geographical analysis, computer-based geographic information systems (GIS).

In their study, geographers use four interrelated approaches:

- Systematic – Groups geographical knowledge into categories that can be explored globally.
- Regional – Examines systematic relationships between categories for a specific region or location on the planet.
- Descriptive – Simply specifies the locations of features and populations.
- Analytical – Asks *why* we find features and populations in a specific geographic area.

Cartography

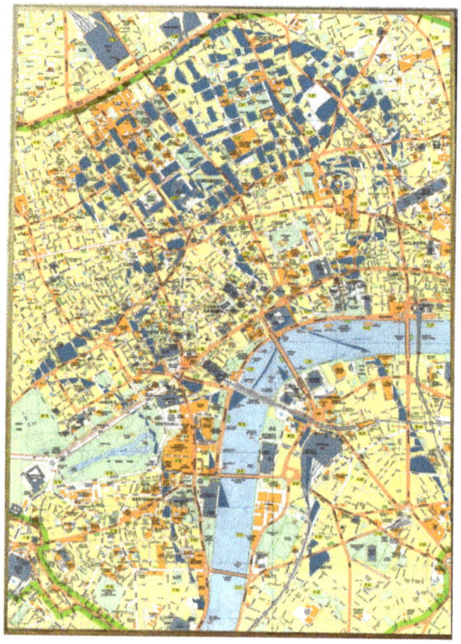

Cartography studies the representation of the Earth's surface with abstract symbols (map making). Although other subdisciplines of geography rely on maps for presenting their analyses, the actual making of maps is abstract enough to be regarded separately. Cartography has grown from a collection of drafting techniques into an actual science.

Cartographers must learn cognitive psychology and ergonomics to understand which symbols convey information about the Earth most effectively, and behavioural psychology to induce the readers of their maps to act on the information. They must learn geodesy and fairly advanced mathematics to understand how the shape of the Earth affects the distortion of map symbols projected onto a flat surface for viewing. It can be said, without much controversy, that cartography is the seed from which the larger field of geography grew. Most geographers will cite a childhood fascination with maps as an early sign they would end up in the field.

Geographic Information Systems

Geographic information systems (GIS) deal with the storage of information about the Earth for automatic retrieval by a computer, in an accurate manner appropriate to the information's purpose. In addition to all of the other subdisciplines of geography, GIS specialists must understand computer science and database systems. GIS has revolutionized the field of cartography: nearly all mapmaking is now done with the assistance of some form of GIS software. GIS also refers to the science of using GIS software and GIS techniques to represent, analyse, and predict the spatial relationships. In this context, GIS stands for geographic information science.

Remote Sensing

Remote sensing is the science of obtaining information about Earth features from measurements made at a distance. Remotely sensed data comes in many forms, such as satellite imagery, aerial photography, and data obtained from hand-held sensors. Geographers increasingly use remotely sensed data to obtain information about the Earth's land surface, ocean, and atmosphere, because it: (a) supplies objective information at a variety of spatial scales (local to global), (b) provides a synoptic view of the area of interest, (c) allows access to distant and inaccessible sites, (d) provides spectral information outside the visible portion of the electromagnetic spectrum, and (e) facilitates studies of how features/areas change over time. Remotely sensed data may be analysed either independently of, or in conjunction with other digital data layers (e.g., in a geographic information system).

Quantitative Methods

Geostatistics deal with quantitative data analysis, specifically the application of statistical methodology to the exploration of geographic phenomena. Geostatistics is used extensively in a variety of fields, including hydrology, geology, petroleum exploration, weather analysis, urban planning, logistics, and epidemiology. The mathematical basis for geostatistics derives from cluster analysis, linear discriminant analysis and non-parametric statistical tests, and a variety of other subjects. Applications of geostatistics rely heavily on geographic information systems, particularly for the interpolation (estimate) of unmeasured points. Geographers are making notable contributions to the method of quantitative techniques.

Qualitative Methods

Geographic qualitative methods, or ethnographical research techniques, are used by human geographers. In cultural geography there is a tradition of employing qualitative research techniques, also used in anthropology and sociology. Participant observation and in-depth interviews provide human geographers with qualitative data.

References

- Earth, place: britannica.com, Retrieved 5 February, 2019
- Atmosphere, science: britannica.com, Retrieved 8 June, 2019
- Biosphere, earth-systems: agci.org, Retrieved 5 March, 2019
- Geosphere, earth-systems: agci.org, Retrieved 11 January, 2019
- Hydrosphere, science: britannica.com, Retrieved 12 July, 2019
- Schlesinger, W.H..; Bernhardt, E.S. (2013). Biogeochemistry: an analysis of global change (3rd ed.). Oxford: Academic Press. ISBN 978-0123858740
- Cryosphere, earth-systems: agci.org, Retrieved 1 August, 2019
- Harper, Douglas. "Online Etymology Dictionary". Etymonline.com. Online Etymology Dictionary. Retrieved 10 November 2016
- Layers-of-the-earth-lies-beneath-earths-crust, trevornace, sites: forbes.com, Retrieved 3 May, 2019
- "Geography". The American Heritage Dictionary/ of the English Language, Fourth Edition. Houghton Mifflin Company. Retrieved 9 October 2006

Geology 2

- Structural Geology
- Hydrogeology
- Paleontology
- Other Branches of Geology
- Plate Tectonics
- Volcanology
- Soil Science
- Applied Geology

The earth science which deals with the rocks that constitute the solid Earth as well as the processes that change them over time is referred to as geology. Some of the major branches of geology are structural geology, hydrogeology, paleontology, petroleum geology and engineering geology. These branches have been thoroughly discussed in this chapter to provide an extensive understanding of the subject.

Geology is the study of Earth—its interior and its exterior surface, the rocks and other materials that are around us, the processes that have resulted in the formation of those materials, the water that flows over the surface and lies underground, the changes that have taken place over the vastness of geological time, and the changes that we can anticipate will take place in the near future. Geology is a science: we use deductive reasoning and scientific methods to understand geological problems.

Geology is arguably the most integrated of all of the sciences because it involves the understanding and application of all of the other sciences: physics, chemistry, biology, mathematics, astronomy, and others. But unlike most of the other sciences, geology has an extra dimension, that of time—billions of years of it. Geologists study the evidence that they see around them, but in most cases, they are observing the results of processes that happened thousands, millions, and even billions of years in the past. Those were processes that took place at incredibly slow rates—millimeters per year to centimeters per year—but because of the amount of time available, they produced massive results.

Rearguard Mountain and Robson Glacier in the Rocky Mountains of British Columbia.

Geology is displayed on a grand scale in mountainous regions, perhaps nowhere better than the Rocky Mountains in Canada. The peak on the right is Rearguard Mountain, which is a few kilometers northeast of Mount Robson, the tallest peak in the Canadian Rockies (3,954 m). The large glacier in the middle of the photo is the Robson Glacier. The river flowing from Robson Glacier drains into Berg Lake in the bottom right.

The sedimentary rock that these mountains are made of formed in ocean water over 500 million years ago. A few hundred million years later, these beds were pushed east for tens to hundreds of kilometers by tectonic plate convergence and also pushed up to thousands of meters above sea level. Over the past two million years this area—like most of the rest of Canada—has been repeatedly glaciated, and the erosional effects of those glaciations are obvious. The Robson Glacier is now only a small remnant of its size during the Little Ice Age of the fifteenth to eighteenth centuries, as shown by the distinctive line on the slope on the left. Like almost all other glaciers in the world, it is now receding even more rapidly because of human-caused climate change.

Geology is also about understanding the evolution of life on Earth; about discovering resources such as metals and energy; about recognizing and minimizing the environmental implications of our use of those resources; and about learning how to mitigate the hazards related to earthquakes, volcanic eruptions, and slope failures.

Structural Geology

Structural geology is the study of the three-dimensional distribution of rock units with respect to their deformational histories. The primary goal of structural geology is to use measurements of present-day rock geometries to uncover information about the history of deformation (strain) in the rocks, and ultimately, to understand the stress field that resulted in the observed strain and geometries. This understanding of the dynamics of the stress field can be linked to important events in the geologic past; a common goal is to understand the structural evolution of a particular area with respect to regionally widespread patterns of rock deformation (e.g., mountain building, rifting) due to plate tectonics.

Use and Importance

The study of geologic structures has been of prime importance in economic geology, both petroleum geology and mining geology. Folded and faulted rock strata commonly form traps that accumulate and concentrate fluids such as petroleum and natural gas. Similarly, faulted and structurally complex areas are notable as permeable zones for hydrothermal fluids, resulting in concentrated areas of base and precious metal ore deposits. Veins of minerals containing various metals commonly occupy faults and fractures in structurally complex areas. These structurally fractured and faulted zones often occur in association with intrusive igneous rocks. They often also occur around geologic reef complexes and collapse features such as ancient sinkholes. Deposits of gold, silver, copper, lead, zinc, and other metals, are commonly located in structurally complex areas.

Structural geology is a critical part of engineering geology, which is concerned with the physical and mechanical properties of natural rocks. Structural fabrics and defects such as faults, folds, foliations and joints are internal weaknesses of rocks which may affect the stability of human engineered structures such as dams, road cuts, open pit mines and underground mines or road tunnels.

Geotechnical risk, including earthquake risk can only be investigated by inspecting a combination of structural geology and geomorphology. In addition, areas of karst landscapes which reside atop underground caverns, potential sinkholes, or other collapse features are of particular importance for these scientists. In addition, areas of steep slopes are potential collapse or landslide hazards.

Environmental geologists and hydrogeologists need to apply the tenets of structural geology to understand how geologic sites impact (or are impacted by) groundwater flow and penetration. For instance, a hydrogeologist may need to determine if seepage of toxic substances from waste dumps is occurring in a residential area or if salty water is seeping into an aquifer.

Plate tectonics is a theory developed during the 1960s which describes the movement of continents by way of the separation and collision of crustal plates. It is in a sense structural geology on a planet scale, and is used throughout structural geology as a framework to analyze and understand global, regional, and local scale features.

Methods

Structural geologists use a variety of methods to (first) measure rock geometries, (second) reconstruct their deformational histories, and (third) estimate the stress field that resulted in that deformation.

Geometries

Primary data sets for structural geology are collected in the field. Structural geologists measure a variety of planar features (bedding planes, foliation planes, fold axial planes, fault planes, and joints), and linear features (stretching lineations, in which minerals are ductily extended; fold axes; and intersection lineations, the trace of a planar feature on another planar surface).

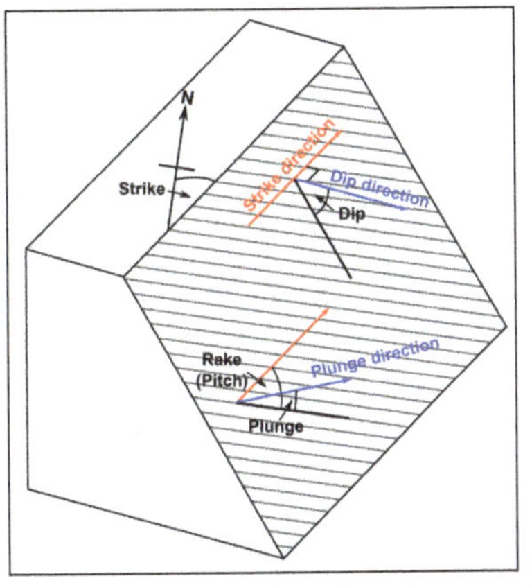

Illustration of measurement conventions for planar and linear structures

Measurement Conventions

The inclination of a planar structure in geology is measured by strike and dip. The strike is the line of intersection between the planar feature and a horizontal plane, taken according to the right hand convention, and the dip is the magnitude of the inclination, below horizontal, at right angles to strike. For example; striking 25 degrees East of North, dipping 45 degrees Southeast, recorded as N25E,45SE.

Alternatively, dip and dip direction may be used as this is absolute. Dip direction is measured in 360 degrees, generally clockwise from North. For example, a dip of 45 degrees towards 115 degrees azimuth, recorded as 45/115. Note that this is the same as above.

The term hade is occasionally used and is the deviation of a plane from vertical i.e. (90°-dip).

Fold axis plunge is measured in dip and dip direction (strictly, plunge and azimuth of plunge). The orientation of a fold axial plane is measured in strike and dip or dip and dip direction.

Lineations are measured in terms of dip and dip direction, if possible. Often lineations occur expressed on a planar surface and can be difficult to measure directly. In this case, the lineation may be measured from the horizontal as a rake or pitch upon the surface.

Rake is measured by placing a protractor flat on the planar surface, with the flat edge horizontal and measuring the angle of the lineation clockwise from horizontal. The orientation of the lineation can then be calculated from the rake and strike-dip information of the plane it was measured from, using a stereographic projection.

If a fault has lineations formed by movement on the plane, e.g.; slickensides, this is recorded as a lineation, with a rake, and annotated as to the indication of throw on the fault.

Generally it is easier to record strike and dip information of planar structures in dip/dip direction format as this will match all the other structural information you may be recording about folds, lineations, etc., although there is an advantage to using different formats that discriminate between planar and linear data.

Plane, Fabric, Fold and Deformation Conventions

The convention for analysing structural geology is to identify the planar structures, often called planar fabrics because this implies a textural formation, the linear structures and, from analysis of these, unravel deformations.

Planar structures are named according to their order of formation, with original sedimentary layering the lowest at So. Often it is impossible to identify So in highly deformed rocks, so numbering may be started at an arbitrary number or given a letter (S_A, for instance). In cases where there is a bedding-plane foliation caused by burial metamorphism or diagenesis this may be enumerated as Soa.

If there are folds, these are numbered as F_1, F_2, etc. Generally the axial plane foliation or cleavage of a fold is created during folding, and the number convention should match. For example, an F_2 fold should have an S_2 axial foliation.

Deformations are numbered according to their order of formation with the letter D denoting a deformation event. For example, D_1, D_2, D_3. Folds and foliations, because they are formed by deformation events, should correlate with these events. For example, an F_2 fold, with an S_2 axial plane foliation would be the result of a D_2 deformation.

Metamorphic events may span multiple deformations. Sometimes it is useful to identify them similarly to the structural features for which they are responsible, e.g.; M_2. This may be possible by observing porphyroblast formation in cleavages of known deformation age, by identifying metamorphic mineral assemblages created by different events, or via geochronology.

Intersection lineations in rocks, as they are the product of the intersection of two planar structures, are named according to the two planar structures from which they are formed. For instance, the intersection lineation of a S_1 cleavage and bedding is the $L_{1\text{-}o}$ intersection lineation (also known as the cleavage-bedding lineation).

Stretching lineations may be difficult to quantify, especially in highly stretched ductile rocks where minimal foliation information is preserved. Where possible, when correlated with deformations (as few are formed in folds, and many are not strictly associated with planar foliations), they may be identified similar to planar surfaces and folds, e.g.; L_1, L_2. For convenience some geologists prefer to annotate them with a subscript S, for example L_{s1} to differentiate them from intersection lineations, though this is generally redundant.

Stereographic Projections

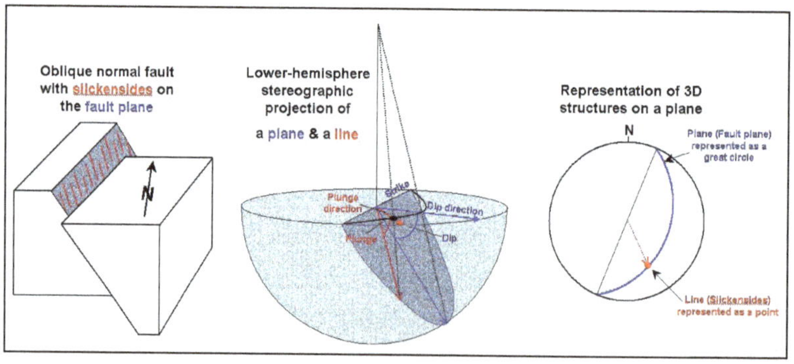

Stereographic projection is a method for analyzing the nature and orientation of deformation stresses, lithological units and penetrative fabrics wherein linear and planar features (structural strike and dip readings, typically taken using a compass clinometer) passing through an imagined sphere are plotted on a two-dimensional grid projection, facilitating more holistic analysis of a set of measurements.

Rock Macro-structures

On a large scale, structural geology is the study of the three-dimensional interaction and relationships of stratigraphic units within terranes of rock or geological regions.

This branch of structural geology deals mainly with the orientation, deformation and relationships of stratigraphy (bedding), which may have been faulted, folded or given a foliation by some tectonic event. This is mainly a geometric science, from which cross sections and three-dimensional block models of rocks, regions, terranes and parts of the Earth's crust can be generated.

Study of regional structure is important in understanding orogeny, plate tectonics and more specifically in the oil, gas and mineral exploration industries as structures such as faults, folds and unconformities are primary controls on ore mineralisation and oil traps.

Modern regional structure is being investigated using seismic tomography and seismic reflection in three dimensions, providing unrivaled images of the Earth's interior, its faults and the deep crust.

Rock Microstructures

Rock microstructure or texture of rocks is studied by structural geologists on a small scale to provide detailed information mainly about metamorphic rocks and some features of sedimentary rocks, most often if they have been folded.

Textural study involves measurement and characterisation of foliations, crenulations, metamorphic minerals, and timing relationships between these structural features and mineralogical features. Usually this involves collection of hand specimens, which may be cut to provide petrographic thin sections which are analysed under a petrographic microscope. Microstructural analysis finds application also in multi-scale statistical analysis, aimed to analyze some rock features showing scale invariance.

Kinematics

Geologists use rock geometry measurements to understand the history of strain in rocks. Strain can take the form of brittle faulting and ductile folding and shearing. Brittle deformation takes place in the shallow crust, and ductile deformation takes place in the deeper crust, where temperatures and pressures are higher.

Stress Fields

By understanding the constitutive relationships between stress and strain in rocks, geologists can translate the observed patterns of rock deformation into a stress field during the geologic past. The following list of features are typically used to determine stress fields from deformational structures.

- In perfectly brittle rocks, faulting occurs at 30° to the greatest compressional stress. (Byerlee's Law)
- The greatest compressive stress is normal to fold axial planes.

Characterization of the Mechanical Properties of Rock

The mechanical properties of rock play a vital role in the structures that form during deformation deep below the earth's crust. The conditions in which a rock is present will result in different structures that geologists observe above ground in the field. The field of structural geology tries to relate the formations that humans see to the changes the rock went through to get to that final structure. Knowing the conditions of deformation that lead to such structures can illuminate the history of the deformation of the rock.

Temperature and pressure play a huge role in the deformation of rock. At the conditions under the earth's crust of extreme high temperature and pressure, rocks are ductile. They can bend, fold or break. Other vital conditions that contribute to the formation of structure of rock under the earth are the stress and strain fields.

Stress-Strain Curve

Stress is a pressure, defined as a directional force over area. When a rock is subjected to stresses, it changes shape. When the stress is released, the rock may or may not return to its original shape. That change in shape is quantified by strain, the change in length over the original length of the material in one dimension. Stress induces strain which ultimately results in a changed structure.

Elastic deformation refers to a reversible deformation. In other words, when stress on the rock is released, the rock returns to its original shape. Reversible, linear, elasticity involves the stretching, compressing, or distortion of atomic bonds. Because there is no breaking of bonds, the material springs back when the force is released. This type of deformation is modeled using a linear relationship between stress and strain, i.e. a Hookean relationship.

$$\in = \frac{\sigma}{E}$$

Where σ denotes stress, $\in$ denotes strain, and E is the elastic modulus, which is material dependent. The elastic modulus is, in effect, a measure of the strength of atomic bonds.

Plastic deformation refers to non-reversible deformation. The relationship between stress and strain for permanent deformation is nonlinear. Stress has caused permanent change of shape in the material by involving the breaking of bonds.

One mechanism of plastic deformation is the movement of dislocations by an applied stress. Because rocks are essentially aggregates of minerals, we can think of them as poly-crystalline materials. Dislocations are a type of crystallographic defect which consists of an extra or missing half plane of atoms in the periodic array of atoms that make up a crystal lattice. Dislocations are present in all real crystallographic materials.

Hardness

Hardness is difficult to quantify. It is a measure of resistance to deformation, specifically permanent deformation. There is precedent for hardness as a surface quality, a measure of the abrasiveness or surface-scratching resistance of a material. If the material being tested, however, is uniform in composition and structure, then the surface of the material is only a few atomic layers thick, and measurements are of the bulk material. Thus, simple surface measurements yield information about the bulk properties. Ways to measure hardness include:

- Mohs Scale
- Dorry abrasion test
- Deval abrasion test
- Indentation hardness

Indentation hardness is used often in metallurgy and materials science and can be thought of as resistance to penetration by an indenter.

Toughness

Toughness can be described best by a material's resistance to cracking. During plastic deformation, a material absorbs energy until fracture occurs. The area under the stress-strain curve is the work required to fracture the material. The toughness modulus is defined as:

$$M_t = \frac{2}{3}\sigma_{UTS} \in_f$$

Where σ_{UTS} is the ultimate tensile strength, and $\in_f$ is the strain at failure. The modulus is the maximum amount of energy per unit volume a material can absorb without fracturing. From the equation for modulus, for large toughness, high strength and high ductility are needed. These two properties are usually mutually exclusive. Brittle materials have low toughness because low plastic deformation decreases the strain (low ductility). Ways to measure toughness include:

- Page impact machine
- Charpy Impact Test

Resilience

Resilience is a measure of the elastic energy absorbed of a material under stress. In other words, the external work performed on a material during deformation. The area under the elastic portion of the stress-strain curve is the strain energy absorbed per unit volume. The resilience modulus is defined as:

$$M_R = \frac{(\sigma_y)^2}{2E}$$

where σ_y is the yield strength of the material and E is the elastic modulus of the material. To increase resilience, one needs increased elastic yield strength and decreased modulus of elasticity.

Hydrogeology

Hydrogeology is the area of geology that deals with the distribution and movement of groundwater in the soil and rocks of the Earth's crust (commonly in aquifers). The terms groundwater hydrology, geohydrology, and hydrogeology are often used interchangeably.

Groundwater engineering, another name for hydrogeology, is a branch of engineering which is concerned with groundwater movement and design of wells, pumps, and drains. The main concerns in groundwater engineering include groundwater contamination, conservation of supplies, and water quality.

Wells are constructed for use in developing nations, as well as for use in developed nations in places which are not connected to a city water system. Wells must be designed and maintained to uphold the integrity of the aquifer, and to prevent contaminants from reaching the groundwater. Controversy arises in the use of groundwater when its usage impacts surface water systems, or when human activity threatens the integrity of the local aquifer system.

Boy drinks from a tap at a NEWAH WASH water project in Puware Shikhar, Udayapur District, Nepal.

Hydrogeology is an interdisciplinary subject; it can be difficult to account fully for the chemical, physical, biological and even legal interactions between soil, water, nature and society. The study of the interaction between groundwater movement and geology can be quite complex. Groundwater does not always follow the surface topography; groundwater follows pressure gradients (flow from high pressure to low), often through fractures and conduits in circuitous paths. Taking into account the interplay of the different facets of a multi-component system often requires knowledge in several diverse fields at both the experimental and theoretical levels. The following is a more traditional introduction to the methods and nomenclature of saturated subsurface hydrology.

Checking wells

Hydrogeology in Relation to Other Fields

Hydrogeology, as stated above, is a branch of the earth sciences dealing with the flow of water through aquifers and other shallow porous media (typically less than 450 m below the land surface). The very shallow flow of water in the subsurface (the upper 3 m) is pertinent to the fields of soil science, agriculture and civil engineering, as well as to hydrogeology. The general flow of fluids (water, hydrocarbons, geothermal fluids, etc.) in deeper formations is also a concern of geologists, geophysicists and petroleum geologists. Groundwater is a slow-moving, viscous fluid (with a Reynolds number less than unity); many of the empirically derived laws of groundwater flow can be alternately derived in fluid mechanics from the special case of Stokes flow (viscosity and pressure terms, but no inertial term).

The mathematical relationships used to describe the flow of water through porous media are the diffusion and Laplace equations, which have applications in many diverse fields. Steady groundwater flow (Laplace equation) has been simulated using electrical, elastic and heat conduction analogies. Transient groundwater flow is analogous to the diffusion of heat in a solid, therefore some solutions to hydrological problems have been adapted from heat transfer literature.

A piezometer is a device used to measure the hydraulic head of groundwater.

Traditionally, the movement of groundwater has been studied separately from surface water, climatology, and even the chemical and microbiological aspects of hydrogeology (the processes are uncoupled). As the field of hydrogeology matures, the strong interactions between groundwater, surface water, water chemistry, soil moisture and even climate are becoming more clear.

California and Washington both require special certification of hydrogeologists to offer professional services to the public. Twenty-nine states require professional licensing for geologists to offer their services to the public, which often includes work within the domains of developing, managing, and/or remediating groundwater resources.

For example: aquifer drawdown or overdrafting and the pumping of fossil water may be a contributing factor to sea-level rise.

Definitions and Material Properties

A water drop.

One of the main tasks a hydrogeologist typically performs is the prediction of future behavior of an aquifer system, based on analysis of past and present observations. Some hypothetical, but characteristic questions asked would be:

- Can the aquifer support another subdivision?
- Will the river dry up if the farmer doubles his irrigation?
- Did the chemicals from the dry cleaning facility travel through the aquifer to my well and make me sick?
- Will the plume of effluent leaving my neighbor's septic system flow to my drinking water well?

Most of these questions can be addressed through simulation of the hydrologic system (using numerical models or analytic equations). Accurate simulation of the aquifer system requires knowledge of the aquifer properties and boundary conditions. Therefore, a common task of the hydrogeologist is determining aquifer properties using aquifer tests.

In order to further characterize aquifers and aquitards some primary and derived physical properties are introduced below. Aquifers are broadly classified as being either confined or unconfined (water table aquifers), and either saturated or unsaturated; the type of aquifer affects what properties control the flow of water in that medium (e.g., the release of water from storage for confined aquifers is related to the storativity, while it is related to the specific yield for unconfined aquifers).

Aquifer

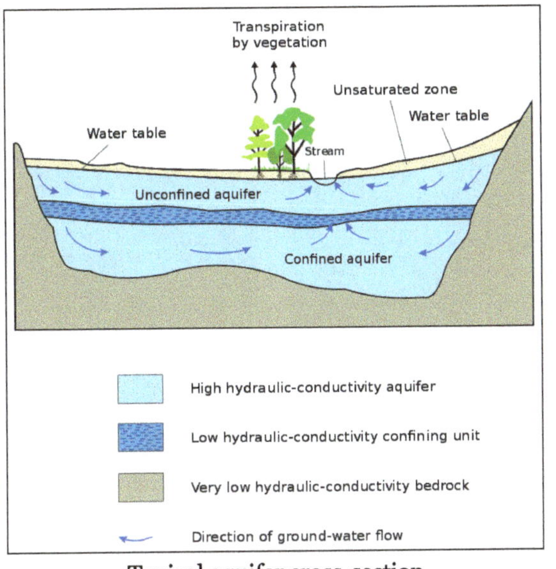

Typical aquifer cross-section

An aquifer is a collection of water underneath the surface, large enough to be useful in a spring or a well. Aquifers can be unconfined, where the top of the aquifer is defined by the water table, or confined, where the aquifer exists underneath a confining bed.

There are three aspects that control the nature of aquifers: stratigraphy, lithology, and geological formations and deposits. The stratigraphy relates the age and geometry of the many formations that compose the aquifer. The lithology refers to the physical components of an aquifer, such as the mineral composition and grain size. The structural features are the elements that arise due to deformations after deposition, such as fractures and folds. Understanding these aspects is paramount to understanding of how an aquifer is formed and how professionals can utilize it for groundwater engineering.

Hydraulic Head

Differences in hydraulic head (h) cause water to move from one place to another; water flows from locations of high h to locations of low h. Hydraulic head is composed of pressure head (ψ) and elevation head (z). The head gradient is the change in hydraulic head per length of flowpath, and appears in Darcy's law as being proportional to the discharge.

Hydraulic head is a directly measurable property that can take on any value (because of the arbitrary datum involved in the z term); ψ can be measured with a pressure transducer (this value can be negative, e.g., suction, but is positive in saturated aquifers), and z can be measured relative to a surveyed datum (typically the top of the well casing). Commonly, in wells tapping unconfined aquifers the water level in a well is used as a proxy for hydraulic head, assuming there is no vertical gradient of pressure. Often only changes in hydraulic head through time are needed, so the constant elevation head term can be left out ($\Delta h = \Delta\psi$).

A record of hydraulic head through time at a well is a hydrograph or, the changes in hydraulic head recorded during the pumping of a well in a test are called drawdown.

Porosity

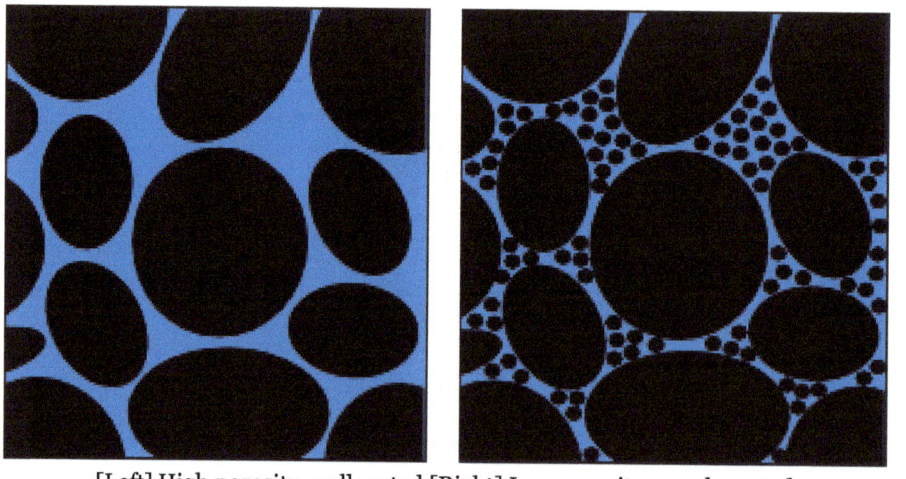

[Left] High porosity, well sorted [Right] Low porosity, poorly sorted

Porosity (n) is a directly measurable aquifer property; it is a fraction between 0 and 1 indicating the amount of pore space between unconsolidated soil particles or within a fractured rock. Typically, the majority of groundwater (and anything dissolved in it) moves through the porosity available to flow (sometimes called effective porosity). Permeability is an expression of the connectedness of the pores. For instance, an unfractured rock unit may have a high porosity (it has lots of holes between its constituent grains), but a low permeability (none of the pores are connected). An example of this phenomenon is pumice, which, when in its unfractured state, can make a poor aquifer.

Porosity does not directly affect the distribution of hydraulic head in an aquifer, but it has a very strong effect on the migration of dissolved contaminants, since it affects groundwater flow velocities through an inversely proportional relationship.

Water Content

Water content (θ) is also a directly measurable property; it is the fraction of the total rock which is filled with liquid water. This is also a fraction between 0 and 1, but it must also be less than or equal to the total porosity.

The water content is very important in vadose zone hydrology, where the hydraulic conductivity is a strongly nonlinear function of water content; this complicates the solution of the unsaturated groundwater flow equation.

Hydraulic Conductivity

Hydraulic conductivity (K) and transmissivity (T) are indirect aquifer properties (they cannot be measured directly). T is the K integrated over the vertical thickness (b) of the aquifer (T=Kb when K is constant over the entire thickness). These properties are measures of an aquifer's ability to transmit water. Intrinsic permeability (κ) is a secondary medium property which does not depend on the viscosity and density of the fluid (K and T are specific to water); it is used more in the petroleum industry.

Specific Storage and Specific Yield

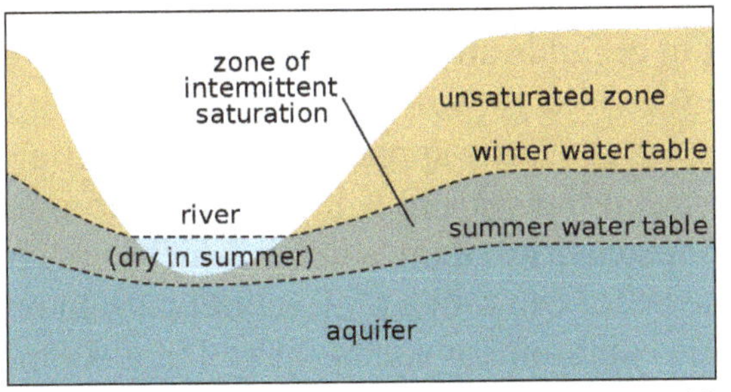

Illustration of seasonal fluctuations in the water table.

Specific storage (S_s) and its depth-integrated equivalent, storativity ($S=S_sb$), are indirect aquifer properties (they cannot be measured directly); they indicate the amount of groundwater released from storage due to a unit depressurization of a confined aquifer. They are fractions between 0 and 1.

Specific yield (S_y) is also a ratio between 0 and 1 ($S_y \leq$ porosity) and indicates the amount of water released due to drainage from lowering the water table in an unconfined aquifer. The value for specific yield is less than the value for porosity because some water will remain in the medium even after drainage due to intermolecular forces. Often the porosity or effective porosity is used as an upper bound to the specific yield. Typically S_y is orders of magnitude larger than S_s.

Contaminant Transport Properties

Often we are interested in how the moving groundwater will transport dissolved contaminants around (the sub-field of contaminant hydrogeology). The contaminants can be man-made (e.g., petroleum products, nitrate, Chromium or radionuclides) or naturally occurring (e.g., arsenic, salinity). Besides needing to understand where the groundwater is flowing, based on the other hydrologic properties discussed above, there are additional aquifer properties which affect how dissolved contaminants move with groundwater.

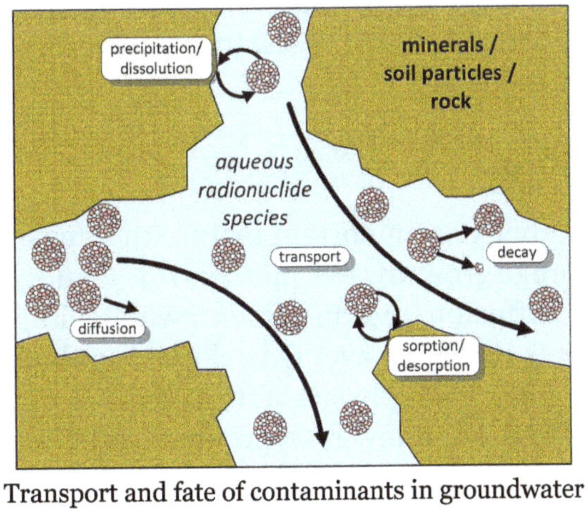

Transport and fate of contaminants in groundwater

Hydrodynamic Dispersion

Hydrodynamic dispersivity (α_L, α_T) is an empirical factor which quantifies how much contaminants stray away from the path of the groundwater which is carrying it. Some of the contaminants will be "behind" or "ahead" the mean groundwater, giving rise to a longitudinal dispersivity (α_L), and some will be "to the sides of" the pure advective groundwater flow, leading to a transverse dispersivity (α_T). Dispersion in groundwater arises because each water "particle", passing beyond a soil particle, must choose where to go, whether left or right or up or down, so that the water "particles" (and their solute) are gradually spread in all directions around the mean path. This is the "microscopic" mechanism, on the scale of soil particles. More important, on long distances, can be the macroscopic inhomogeneities of the aquifer, which can have regions of larger or smaller permeability, so that some water can find a preferential path in one direction, some other in a different direction, so that the contaminant can be spread in a completely irregular way, like in a (three-dimensional) delta of a river.

Dispersivity is actually a factor which represents our lack of information about the system we are simulating. There are many small details about the aquifer which are being averaged when using a macroscopic approach (e.g., tiny beds of gravel and clay in sand aquifers), they manifest themselves as an apparent dispersivity. Because of this, α is often claimed to be dependent on the length scale of the problem — the dispersivity found for transport through 1 m^3 of aquifer is different from that for transport through 1 cm^3 of the same aquifer material.

Molecular Diffusion

Diffusion is a fundamental physical phenomenon, which Einstein characterized as Brownian motion, that describes the random thermal movement of molecules and small particles in gases and liquids. It is an important phenomenon for small distances (it is essential for the achievement of thermodynamic equilibria), but, as the time necessary to cover a distance by diffusion is proportional to the square of the distance itself, it is ineffective for spreading a solute over macroscopic distances. The diffusion coefficient, D, is typically quite small, and its effect can often be considered negligible (unless groundwater flow velocities are extremely low, as they are in clay aquitards).

It is important not to confuse diffusion with dispersion, as the former is a physical phenomenon and the latter is an empirical factor which is cast into a similar form as diffusion, because we already know how to solve that problem.

Retardation by Adsorption

The retardation factor is another very important feature that make the motion of the contaminant to deviate from the average groundwater motion. It is analogous to the retardation factor of chromatography. Unlike diffusion and dispersion, which simply spread the contaminant, the retardation factor changes its global average velocity, so that it can be much slower than that of water. This is due to a chemico-physical effect: the adsorption to the soil, which holds the contaminant back and does not allow it to progress until the quantity corresponding to the chemical adsorption equilibrium has been adsorbed. This effect is particularly important for less soluble contaminants, which thus can move even hundreds or thousands times slower than water. The effect of this phenomenon is that only more soluble species can cover long distances. The retardation factor depends on the chemical nature of both the contaminant and the aquifer.

Governing Equations

Darcy's Law

Darcy's law is a constitutive equation, empirically derived by Henry Darcy in 1856, which states that the amount of groundwater discharging through a given portion of aquifer is proportional to the cross-sectional area of flow, the hydraulic gradient, and the hydraulic conductivity.

Groundwater Flow Equation

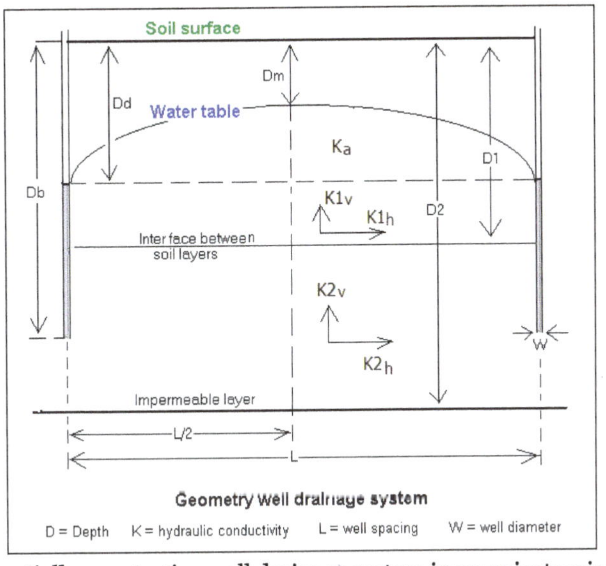

Geometry of a partially penetrating well drainage system in an anisotropic layered aquifer

The groundwater flow equation, in its most general form, describes the movement of groundwater in a porous medium (aquifers and aquitards). It is known in mathematics as the diffusion equation, and has many analogs in other fields. Many solutions for groundwater flow problems were borrowed or adapted from existing heat transfer solutions.

It is often derived from a physical basis using Darcy's law and a conservation of mass for a small control volume. The equation is often used to predict flow to wells, which have radial symmetry, so the flow equation is commonly solved in polar or cylindrical coordinates.

The Theis equation is one of the most commonly used and fundamental solutions to the groundwater flow equation; it can be used to predict the transient evolution of head due to the effects of pumping one or a number of pumping wells.

The Thiem equation is a solution to the steady state groundwater flow equation (Laplace's Equation) for flow to a well. Unless there are large sources of water nearby (a river or lake), true steady-state is rarely achieved in reality.

Both above equations are used in aquifer tests (pump tests).

The Hooghoudt equation is a groundwater flow equation applied to subsurface drainage by pipes, tile drains or ditches. An alternative subsurface drainage method is drainage by wells for which groundwater flow equations are also available.

Calculation of Groundwater Flow

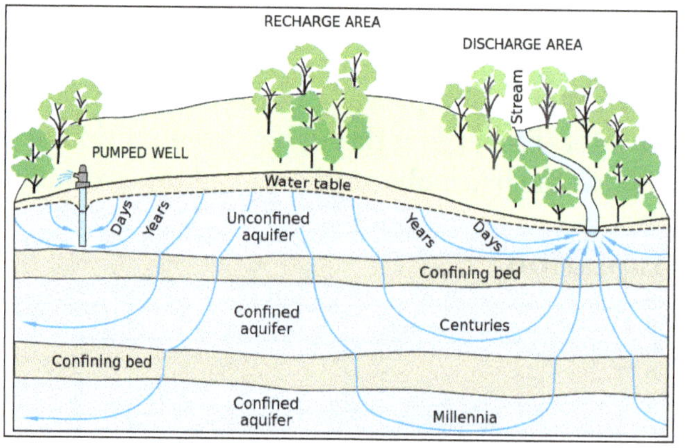

Relative groundwater travel times.

To use the groundwater flow equation to estimate the distribution of hydraulic heads, or the direction and rate of groundwater flow, this partial differential equation (PDE) must be solved. The most common means of analytically solving the diffusion equation in the hydrogeology literature are:

- Laplace, Hankel and Fourier transforms (to reduce the number of dimensions of the PDE),
- similarity transform (also called the Boltzmann transform) is commonly how the Theis solution is derived,
- separation of variables, which is more useful for non-Cartesian coordinates, and
- Green's functions, which is another common method for deriving the Theis solution – from the fundamental solution to the diffusion equation in free space.

No matter which method we use to solve the groundwater flow equation, we need both initial conditions (heads at time (t) = 0) and boundary conditions (representing either the physical boundaries of the domain, or an approximation of the domain beyond that point). Often the initial conditions are supplied to a transient simulation, by a corresponding steady-state simulation (where the time derivative in the groundwater flow equation is set equal to 0).

There are two broad categories of how the (PDE) would be solved; either analytical methods, numerical methods, or something possibly in between. Typically, analytic methods solve the groundwater flow equation under a simplified set of conditions exactly, while numerical methods solve it under more general conditions to an approximation.

Analytic Methods

Analytic methods typically use the structure of mathematics to arrive at a simple, elegant solution, but the required derivation for all but the simplest domain geometries can be quite complex (involving non-standard coordinates, conformal mapping, etc.). Analytic solutions typically are also simply an equation that can give a quick answer based on a few basic parameters. The Theis equation is a very simple (yet still very useful) analytic solution to the groundwater flow equation, typically used to analyze the results of an aquifer test or slug test.

Numerical Methods

The topic of numerical methods is quite large, obviously being of use to most fields of engineering and science in general. Numerical methods have been around much longer than computers have (In the 1920s Richardson developed some of the finite difference schemes still in use today, but they were calculated by hand, using paper and pencil, by human "calculators"), but they have become very important through the availability of fast and cheap personal computers. A quick survey of the main numerical methods used in hydrogeology.

There are two broad categories of numerical methods: gridded or discretized methods and non-gridded or mesh-free methods. In the common finite difference method and finite element method (FEM) the domain is completely gridded ("cut" into a grid or mesh of small elements). The analytic element method (AEM) and the boundary integral equation method (BIEM — sometimes also called BEM, or Boundary Element Method) are only discretized at boundaries or along flow elements (line sinks, area sources, etc.), the majority of the domain is mesh-free.

General Properties of Gridded Methods

Gridded Methods like finite difference and finite element methods solve the groundwater flow equation by breaking the problem area (domain) into many small elements (squares, rectangles, triangles, blocks, tetrahedra, etc.) and solving the flow equation for each element (all material properties are assumed constant or possibly linearly variable within an element), then linking together all the elements using conservation of mass across the boundaries between the elements (similar to the divergence theorem). This results in a system which overall approximates the groundwater flow equation, but exactly matches the boundary conditions (the head or flux is specified in the elements which intersect the boundaries).

Finite differences are a way of representing continuous differential operators using discrete intervals (Δx and Δt), and the finite difference methods are based on these (they are derived from a Taylor series). For example, the first-order time derivative is often approximated using the following forward finite difference, where the subscripts indicate a discrete time location,

$$\frac{\partial h}{\partial t} = h'(t_i) \approx \frac{h_i - h_{i-1}}{\Delta t}.$$

The forward finite difference approximation is unconditionally stable, but leads to an implicit set of equations (that must be solved using matrix methods, e.g. LU or Cholesky decomposition). The similar backwards difference is only conditionally stable, but it is explicit and can be used to "march" forward in the time direction, solving one grid node at a time (or possibly in parallel, since one node depends only on its immediate neighbors). Rather than the finite difference method, sometimes the Galerkin FEM approximation is used in space (this is different from the type of FEM often used in structural engineering) with finite differences still used in time.

Application of Finite Difference Models

MODFLOW is a well-known example of a general finite difference groundwater flow model. It is developed by the US Geological Survey as a modular and extensible simulation tool for modeling groundwater flow. It is free software developed, documented and distributed by the USGS. Many commercial products have grown up around it, providing graphical user interfaces to its input file based interface, and typically incorporating pre- and post-processing of user data. Many other models have been developed to work with MODFLOW input and output, making linked models which simulate several hydrologic processes possible (flow and transport models, surface water and groundwater models and chemical reaction models), because of the simple, well documented nature of MODFLOW.

Application of Finite Element Models

Finite Element programs are more flexible in design (triangular elements vs. the block elements most finite difference models use) and there are some programs available (SUTRA, a 2D or 3D density-dependent flow model by the USGS; Hydrus, a commercial unsaturated flow model; FEFLOW, a commercial modelling environment for subsurface flow, solute and heat transport processes; OpenGeoSys, a scientific open-source project for thermo-hydro-mechanical-chemical (THMC) processes in porous and fractured media; COMSOL Multiphysics (a commercial general modelling environment), FEATool Multiphysics an easy to use MATLAB simulation toolbox, and Integrated Water Flow Model (IWFM), but they are still not as popular in with practicing hydrogeologists as MODFLOW is. Finite element models are more popular in university and laboratory environments, where specialized models solve non-standard forms of the flow equation (unsaturated flow, density dependent flow, coupled heat and groundwater flow, etc.)

Application of Finite Volume Models

The finite volume method is a method for representing and evaluating partial differential equations as algebraic equations. Similar to the finite difference method, values are calculated at discrete places on a meshed geometry. "Finite volume" refers to the small volume surrounding each node point on a mesh. In the finite volume method, volume integrals in a partial differential equation that contain a divergence term are converted to surface integrals, using the divergence theorem. These terms are then evaluated as fluxes at the surfaces of each finite volume. Because the flux entering a given volume is identical to that leaving the adjacent volume, these methods are conservative. Another advantage of the finite volume method is that it is easily formulated to allow for unstructured meshes. The method is used in many computational fluid dynamics packages.

PORFLOW software package is a comprehensive mathematical model for simulation of Ground Water Flow and Nuclear Waste Management developed by Analytic & Computational Research, Inc., ACRi.

The FEHM software package is available free from Los Alamos National Laboratory. This versatile porous flow simulator includes capabilities to model multiphase, thermal, stress, and multicomponent reactive chemistry. Current work using this code includes simulation of methane hydrate formation, CO_2 sequestration, oil shale extraction, migration of both nuclear and chemical contaminants, environmental isotope migration in the unsaturated zone, and karst formation.

Other Methods

These include mesh-free methods like the Analytic Element Method (AEM) and the Boundary Element Method (BEM), which are closer to analytic solutions, but they do approximate the groundwater flow equation in some way. The BEM and AEM exactly solve the groundwater flow equation (perfect mass balance), while approximating the boundary conditions. These methods are more exact and can be much more elegant solutions (like analytic methods are), but have not seen as widespread use outside academic and research groups yet.

Water Wells

A water well is a mechanism for bringing groundwater to the surface by drilling or digging and bringing it up to the surface with a pump or by hand using buckets or similar devices. The first historical instance of water wells was in the 52nd century BC in modern-day Austria. Today, wells are used all over the world, from developing nations to suburbs in the United States.

There are three main types of wells, shallow, deep, and artesian. Shallow wells tap into unconfined aquifers, and are, generally, shallow, less than 15 meters deep. Shallow wells have a small diameter, usually less than 15 centimeters. Deep wells access confined aquifers, and are always drilled by machine. All deep wells bring water to the surface using mechanical pumps. In artesian wells, water flows naturally without the use of a pump or some other mechanical device. This is due the top of the well being located below the water table.

Water Well Design and Construction

A water well in Kerala, India.

One of the most important aspects of groundwater engineering and hydrogeology is water well design and construction. Proper well design and construction are important to maintain the health of the groundwater and the people which will use the well. Factors which must be considered in well design are:

- A reliable aquifer, providing a continuous water supply
- The quality of the accessible groundwater
- How to monitor the well
- Operating costs of the well

- Expected yield of the well
- Any prior drilling into the aquifer

There are five main areas to be considered when planning and constructing a new water well, along with the factors above. They are:

- Aquifer Suitability
- Well Design Considerations
- Well Drilling Methods
- Well Screen Design and Development
- Well Testing

Aquifer suitability starts with determining possible locations for the well using "USGS reports, well logs, and cross sections" of the aquifer. This information should be used to determine aquifer properties such as depth, thickness, transmissivity, and well yield. In this stage, the quality of the water in the aquifer should also be determined, and screening should occur to check for contaminants.

After factors such as depth and well yield are determined, the well design and drilling approach must be established. Drilling method is selected based on "soil conditions, well depth, design, and costs." At this stage, cost estimates are prepared, and plans are adjusted to meet budgetary needs.

Important parts of a well include the well seals, casings or liners, drive shoes, well screen assemblies, and a sand or gravel pack (optional). Each of these components ensures that the well only draws from one aquifer, and no leakage occurs at any stage of the process.

There are several methods of drilling which can be used when constructing a water well. They include: "Cable tool, Air rotary, Mud rotary, and Flooded reverse circulation dual rotary" drilling techniques. Cable tool drilling is inexpensive and can be used for all types of wells, but the alignment must be constantly checked and it has a slow advance rate. It is not an effective drilling technique for consolidated formations, but does provide a small drilling footprint. Air rotary drilling is cost effective and works well for consolidated formations. It has a fast advance rate, but is not adequate for large diameter wells. Mud rotary drilling is especially cost effective for deep wells. It maintains good alignment, but requires a larger footprint. It has a very fast advance rate. Flooded reverse circulation dual rotary drilling is more expensive, but good for large well designs. It is versatile and maintains alignment. It has a fast advance rate.

Well screens ensure that only water makes it to the surface, and sediments remain beneath the Earth's surface. Screens are placed along the shaft of the well to filter out sediment as water is pumped towards the surface. Screen design can be impacted by the nature of the soil, and natural pack designs can be used to maximize efficiency.

After construction of the well, testing must be done to assess productivity, efficiency and yield of the well, as well as determine the impacts of the well on the aquifer. Several different tests should be completed on the well in order to test all relevant qualities of the well.

Issues in Groundwater Engineering and Hydrogeology

Contamination

Groundwater contamination happens when other fluids seep into the aquifer and mix with existing groundwater. Pesticides, fertilizers, and gasoline are common contaminants of aquifers. Underground storage tanks for chemicals such as gasoline are especially concerning sources of groundwater contamination. As these tanks corrode, they can leak, and their contents can contaminate nearby groundwater. For buildings which are not connected to a wastewater treatment system, septic tanks can be used to dispose of waste at a safe rate. If septic tanks are not built or maintained properly, they can leak bacteria, viruses and other chemicals into the surrounding groundwater. Landfills are another potential source of groundwater contamination. As trash is buried, harmful chemicals can migrate from the garbage and into the surrounding groundwater if the protective base layer is cracked or otherwise damaged. Other chemicals, such as road salts and chemicals used on lawns and farms, can runoff into local reservoirs, and eventually into aquifers. As water goes through the water cycle, contaminants in the atmosphere can contaminate the water. This water can also make its way into groundwater.

Controversy

Fracking

Contamination of groundwater due to fracking has long been debated. Since chemicals commonly used in hydraulic fracturing are not tested by government agencies responsible for determining the effects of fracking on groundwater, laboratories at the United States Environmental Protection Agency, or EPA, have a hard time determining if chemicals used in fracking are present in nearby aquifers. In 2016, the EPA released a report which states that drinking water can be contaminated by fracking. This was a reversal of their previous policies after a $29 million study into the effects of fracking on local drinking water.

California

California sees some of the largest controversies in groundwater usage due to the dry conditions California faces, high population, and intensive agriculture. Conflicts generally occur over pumping groundwater and shipping it out of the area, unfair use of water by a commercial company, and contamination of groundwater by development projects. In Siskiyou County in northern California, the California Superior Court ruled poor groundwater regulations have allowed pumping to diminish the flows in the Scott River and disturbed the natural habitat of salmon. In Owens Valley in central California, groundwater was pumped for use in fish farms, which resulted in the death of local meadows and other ecosystems. This resulted in a lawsuit and settlement against the fish companies. Development in southern California is threatening local aquifers, contaminating groundwater through construction and normal human activity. For example, a solar project in San Bernardino County would allegedly threaten the ecosystem of bird and wildlife species because of its use of up to 1.3 million cubic meters of groundwater, which could impact Harper Lake. In September 2014, California passed the Sustainable Groundwater Management Act, which requires users to manage groundwater appropriately, as it is connected to surface water systems.

Colorado

Due to its arid climate, the state of Colorado gets most of its water from underground. Because of this, there have been issues regarding groundwater engineering practices. As many as 65,000 people were affected when high levels of PFCs were found in the Widefield Aquifer. Groundwater use in Colorado dates back to before the 20th century. Nineteen of Colorado's 63 counties depend mostly on groundwater for supplies and domestic uses. The Colorado Geological Survey has three significant reports on groundwater in the Denver Basin. The first report Geology of Upper Cretaceous, Paleocene and Eocene Strata in the Southwestern Denver Basin, The second report Bedrock Geology, Structure, and Isopach Maps of the Upper Cretaceous to Paleogene Strata between Greeley and Colorado Springs, The third publication Cross Sections of the Freshwater Bearing Strata of the Denver Basin between Greeley and Colorado Springs.

New Trends in Groundwater

Engineering/Hydrogeology

Since the first wells were made thousands of years ago, groundwater systems have been changed by human activity. Fifty years ago, the sustainability of these systems on a larger scale began to come into consideration, becoming one of the main focuses of groundwater engineering. New ideas and research are advancing groundwater engineering into the 21st century, while still considering groundwater conservation.

Topographical Mapping

New advancements have arisen in topographical mapping to improve sustainability. Topographic mapping has been updated to include radar, which can penetrate the ground to help pinpoint areas of concern. In addition, large computations can use gathered data from maps to further the knowledge of groundwater aquifers in recent years. This has made highly complex and individualized water cycle models possible, which has helped to make groundwater sustainability more applicable to specific situations.

The Role of Technology

Technological improvements have advanced topographical mapping, and have also improved the quality of lithosphere, hydrosphere, biosphere, and atmosphere simulations. These simulations are useful on their own; however, when used together, they help to give an even more accurate prediction of the future sustainability of an area, and what changes can be made to ensure stability in the area. This would not be possible without the advancement of technology. As technology continues to progress, the simulations will increase in accuracy and allow for more complex studies and projects in groundwater engineering.

Growing Populations

As populations continue to grow, areas which were using groundwater at a sustainable rate are now beginning to face sustainability issues for the future. Populations of the size currently seen in large cities were not taken into consideration when the long term sustainability of aquifers. These large population sizes are beginning to stress groundwater supply. This has led to the need for new policies in some urban areas. These are known as proactive land-use management, where cities can move proactively to conserve groundwater.

In Brazil, overpopulation caused municipally provided water to run low. Due to the shortage of water, people began to drill wells within the range normally served by the municipal water system. This was a solution for people in high socioeconomic standing, but left much of the underprivileged population without access to water. Because of this, a new municipal policy was created which drilled wells to assist those who could not afford to drill wells of their own. Because the city is in charge of drilling the new wells, they can better plan for the future sustainability of the groundwater in the region, by carefully placing the wells and taking growing populations into consideration.

Dependency on Groundwater in the United States

In the United States, 51% of the drinking water comes from groundwater supplies. Around 99% of the rural population depends on groundwater. In addition, 64% of the total groundwater of the country is used for irrigation, and some of it is used for industrial processes and recharge for lakes and rivers. In 2010, 22 percent of freshwater used in USA came from groundwater and the other 78 percent came from surface water. Groundwater is important for some states that don't have access to fresh water. most of the fresh groundwater 65 percent is used for irrigation and the 21 percent is used for public purposes drinking mostly.

Paleontology

A paleontologist at work at John Day Fossil Beds National Monument.

Paleontology, sometimes spelled palaeontology is the scientific study of life that existed prior to, and sometimes including, the start of the Holocene Epoch (roughly 11,700 years before present). It includes the study of fossils to determine organisms' evolution and interactions with each other and their environments (their paleoecology). Paleontological observations have been documented as far back as the 5th century BC. The science became established in the 18th century as a result of Georges Cuvier's work on comparative anatomy, and developed rapidly in the 19th century.

Paleontology lies on the border between biology and geology, but differs from archaeology in that it excludes the study of anatomically modern humans. It now uses techniques drawn from a wide range of sciences, including biochemistry, mathematics, and engineering. Use of all these techniques has enabled paleontologists to discover much of the evolutionary history of life, almost all the way back to when Earth became capable of supporting life, about 3.8 billion years ago. As knowledge has increased, paleontology has developed specialised sub-divisions, some of which focus on different types of fossil organisms while others study ecology and environmental history, such as ancient climates.

Body fossils and trace fossils are the principal types of evidence about ancient life, and geochemical evidence has helped to decipher the evolution of life before there were organisms large enough to leave body fossils. Estimating the dates of these remains is essential but difficult: sometimes adjacent rock layers allow radiometric dating, which provides absolute dates that are accurate to within 0.5%, but more often paleontologists have to rely on relative dating by solving the "jigsaw puzzles" of biostratigraphy (arrangement of rock layers from youngest to oldest). Classifying ancient organisms is also difficult, as many do not fit well into the Linnaean taxonomy classifying living organisms, and paleontologists more often use cladistics to draw up evolutionary "family trees". The final quarter of the 20th century saw the development of molecular phylogenetics, which investigates how closely organisms are related by measuring the similarity of the DNA in their genomes. Molecular phylogenetics has also been used to estimate the dates when species diverged, but there is controversy about the reliability of the molecular clock on which such estimates depend.

The simplest definition of paleontology is "the study of ancient life". The field seeks information about several aspects of past organisms: "their identity and origin, their environment and evolution, and what they can tell us about the Earth's organic and inorganic past".

A Historical Science

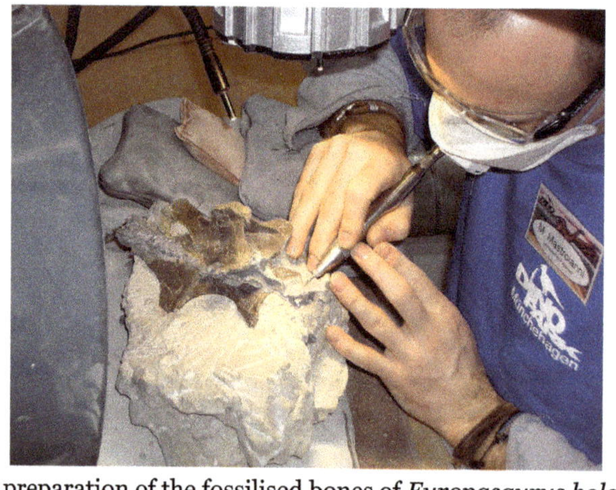

The preparation of the fossilised bones of *Europasaurus holgeri*.

Paleontology is one of the historical sciences, along with archaeology, geology, astronomy, cosmology, philology and history itself: it aims to describe phenomena of the past and reconstruct their causes. Hence it has three main elements: description of past phenomena; developing a general theory about the causes of various types of change; and applying those theories to specific facts.

When trying to explain the past, paleontologists and other historical scientists often construct a set of hypotheses about the causes and then look for a *smoking gun*, a piece of evidence that strongly accords with one hypothesis over the others. Sometimes the smoking gun is discovered by a fortunate accident during other research. For example, the discovery by Luis and Walter Alvarez of iridium, a mainly extraterrestrial metal, in the Cretaceous–Tertiary boundary layer made asteroid impact the most favored explanation for the Cretaceous–Paleogene extinction event, although the contribution of volcanism continues to be debated.

The other main type of science is experimental science, which is often said to work by conducting experiments to *disprove* hypotheses about the workings and causes of natural phenomena. This approach cannot prove a hypothesis, since some later experiment may disprove it, but the accumulation of failures to disprove is often compelling evidence in favor. However, when confronted with totally unexpected phenomena, such as the first evidence for invisible radiation, experimental scientists often use the same approach as historical scientists: construct a set of hypotheses about the causes and then look for a "smoking gun".

Paleontology lies between biology and geology since it focuses on the record of past life, but its main source of evidence is fossils in rocks. For historical reasons, paleontology is part of the geology department at many universities: in the 19th and early 20th centuries, geology departments found fossil evidence important for dating rocks, while biology departments showed little interest.

Paleontology also has some overlap with archaeology, which primarily works with objects made by humans and with human remains, while paleontologists are interested in the characteristics and evolution of humans as a species. When dealing with evidence about humans, archaeologists and paleontologists may work together – for example paleontologists might identify animal or plant fossils around an archaeological site, to discover what the people who lived there ate; or they might analyze the climate at the time of habitation.

In addition, paleontology often borrows techniques from other sciences, including biology, osteology, ecology, chemistry, physics and mathematics. For example, geochemical signatures from rocks may help to discover when life first arose on Earth, and analyses of carbon isotope ratios may help to identify climate changes and even to explain major transitions such as the Permian–Triassic extinction event. A relatively recent discipline, molecular phylogenetics, compares the DNA and RNA of modern organisms to re-construct the "family trees" of their evolutionary ancestors. It has also been used to estimate the dates of important evolutionary developments, although this approach is controversial because of doubts about the reliability of the "molecular clock". Techniques from engineering have been used to analyse how the bodies of ancient organisms might have worked, for example the running speed and bite strength of *Tyrannosaurus*, or the flight mechanics of *Microraptor*. It is relatively commonplace to study the internal details of fossils using X-ray microtomography. Paleontology, biology, archaeology, and paleoneurobiology combine to study endocranial casts (endocasts) of species related to humans to clarify the evolution of the human brain.

Paleontology even contributes to astrobiology, the investigation of possible life on other planets, by developing models of how life may have arisen and by providing techniques for detecting evidence of life.

Subdivisions

As knowledge has increased, paleontology has developed specialised subdivisions. Vertebrate paleontology concentrates on fossils from the earliest fish to the immediate ancestors of modern mammals. Invertebrate paleontology deals with fossils such as molluscs, arthropods, annelid worms and echinoderms. Paleobotany studies fossil plants, algae, and fungi. Palynology, the study of pollen and spores produced by land plants and protists, straddles paleontology and botany, as it deals with both living and fossil organisms. Micropaleontology deals with microscopic fossil organisms of all kinds.

Analyses using engineering techniques show that *Tyrannosaurus* had a devastating bite, but raise doubts about its running ability.

Instead of focusing on individual organisms, paleoecology examines the interactions between different ancient organisms, such as their food chains, and the two-way interactions with their environments. For example, the development of oxygenic photosynthesis by bacteria caused the oxygenation of the atmosphere and hugely increased the productivity and diversity of ecosystems. Together, these led to the evolution of complex eukaryotic cells, from which all multicellular organisms are built.

Paleoclimatology, although sometimes treated as part of paleoecology, focuses more on the history of Earth's climate and the mechanisms that have changed it – which have sometimes included evolutionary developments, for example the rapid expansion of land plants in the Devonian period removed more carbon dioxide from the atmosphere, reducing the greenhouse effect and thus helping to cause an ice age in the Carboniferous period.

Biostratigraphy, the use of fossils to work out the chronological order in which rocks were formed, is useful to both paleontologists and geologists. Biogeography studies the spatial distribution of organisms, and is also linked to geology, which explains how Earth's geography has changed over time.

Sources of Evidence

Body Fossils

Fossils of organisms' bodies are usually the most informative type of evidence. The most common types are wood, bones, and shells. Fossilisation is a rare event, and most fossils are destroyed by erosion or metamorphism before they can be observed. Hence the fossil record is very incomplete, increasingly so further back in time. Despite this, it is often adequate to illustrate the broader patterns of life's history. There are also biases in the fossil record: different environments are more favorable to the preservation of different types of organism or parts of organisms. Further, only the parts of organisms that were already mineralised are usually preserved, such as the shells of molluscs. Since most animal species are soft-bodied, they decay before they can become fossilised. As a result, although there are 30-plus phyla of living animals, two-thirds have never been found as fossils.

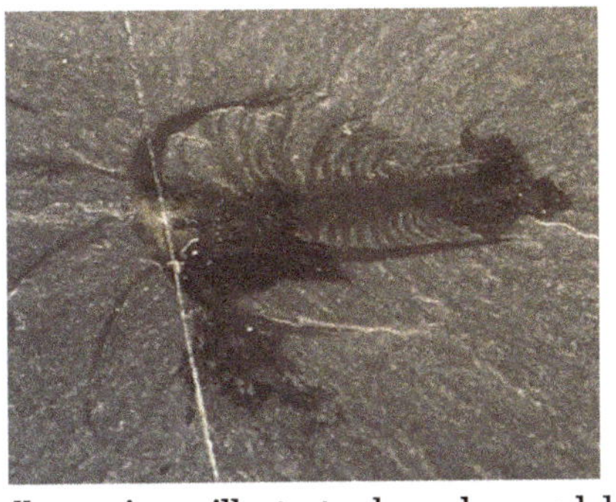

This *Marrella* specimen illustrates how clear and detailed the fossils from the Burgess Shale lagerstätte are.

Occasionally, unusual environments may preserve soft tissues. These lagerstätten allow paleontologists to examine the internal anatomy of animals that in other sediments are represented only by shells, spines, claws, etc. – if they are preserved at all. However, even lagerstätten present an incomplete picture of life at the time. The majority of organisms living at the time are probably not represented because lagerstätten are restricted to a narrow range of environments, e.g. where soft-bodied organisms can be preserved very quickly by events such as mudslides; and the exceptional events that cause quick burial make it difficult to study the normal environments of the animals. The sparseness of the fossil record means that organisms are expected to exist long before and after they are found in the fossil record – this is known as the Signor–Lipps effect.

Trace Fossils

Cambrian trace fossils including *Rusophycus*, made by a trilobite.

Trace fossils consist mainly of tracks and burrows, but also include coprolites (fossil feces) and marks left by feeding. Trace fossils are particularly significant because they represent a data source that is not limited to animals with easily fossilised hard parts, and they reflect organisms' behaviours. Also many traces date from significantly earlier than the body fossils of animals that are thought to have been capable of making them. Whilst exact assignment of trace fossils to their makers is generally impossible, traces may for example provide the earliest physical evidence of the appearance of moderately complex animals (comparable to earthworms).

Geochemical Observations

Geochemical observations may help to deduce the global level of biological activity at a certain period, or the affinity of certain fossils. For example, geochemical features of rocks may reveal when life first arose on Earth, and may provide evidence of the presence of eukaryotic cells, the type from which all multicellular organisms are built. Analyses of carbon isotope ratios may help to explain major transitions such as the Permian–Triassic extinction event.

Classifying Ancient Organisms

Naming groups of organisms in a way that is clear and widely agreed is important, as some disputes in paleontology have been based just on misunderstandings over names. Linnaean taxonomy is commonly used for classifying living organisms, but runs into difficulties when dealing with newly discovered organisms that are significantly different from known ones. For example: it is hard to decide at what level to place a new higher-level grouping, e.g. genus or family or order; this is important since the Linnaean rules for naming groups are tied to their levels, and hence if a group is moved to a different level it must be renamed.

Paleontologists generally use approaches based on cladistics, a technique for working out the evolutionary "family tree" of a set of organisms. It works by the logic that, if groups B and C have more similarities to each other than either has to group A, then B and C are more closely related to each other than either is to A. Characters that are compared may be anatomical, such as the presence of a notochord, or molecular, by comparing sequences of DNA or proteins. The result of a successful analysis is a hierarchy of clades – groups that share a common ancestor. Ideally the "family tree" has only two branches leading from each node ("junction"), but sometimes there is too little information to achieve this and paleontologists have to make do with junctions that have several branches. The cladistic technique is sometimes fallible, as some features, such as wings or camera eyes, evolved more than once, convergently – this must be taken into account in analyses.

Evolutionary developmental biology, commonly abbreviated to "Evo Devo", also helps paleontologists to produce "family trees", and understand fossils. For example, the embryological development of some modern brachiopods suggests that brachiopods may be descendants of the halkieriids, which became extinct in the Cambrian period.

Estimating the Dates of Organisms

Paleontology seeks to map out how living things have changed through time. A substantial hurdle to this aim is the difficulty of working out how old fossils are. Beds that preserve fossils typically lack the radioactive elements needed for radiometric dating. This technique is our only means of giving rocks greater than about 50 million years old an absolute age, and can be accurate to within 0.5% or better. Although radiometric dating requires very careful laboratory work, its basic principle is simple: the rates at which various radioactive elements decay are known, and so the ratio of the radioactive element to the element into which it decays shows how long ago the radioactive element was incorporated into the rock. Radioactive elements are common only in rocks with a volcanic origin, and so the only fossil-bearing rocks that can be dated radiometrically are a few volcanic ash layers.

Consequently, paleontologists must usually rely on stratigraphy to date fossils. Stratigraphy is the science of deciphering the "layer-cake" that is the sedimentary record, and has been compared to a jigsaw puzzle. Rocks normally form relatively horizontal layers, with each layer younger than the one underneath it. If a fossil is found between two layers whose ages are known, the fossil's age must lie between the two known ages. Because rock sequences are not continuous, but may be broken up by faults or periods of erosion, it is very difficult to match up rock beds that are not directly next to one another. However, fossils of species that survived for a relatively short time can be used to link up isolated rocks: this technique is called *biostratigraphy*. For instance, the conodont *Eoplacognathus pseudoplanus* has a short range in the Middle Ordovician period. If rocks of unknown age are found to have traces of *E. pseudoplanus*, they must have a mid-Ordovician age. Such index fossils must be distinctive, be globally distributed and have a short time range to be useful. However, misleading results are produced if the index fossils turn out to have longer fossil ranges than first thought. Stratigraphy and biostratigraphy can in general provide only relative dating (*A* was before *B*), which is often sufficient for studying evolution. However, this is difficult for some time periods, because of the problems involved in matching up rocks of the same age across different continents.

Family-tree relationships may also help to narrow down the date when lineages first appeared. For instance, if fossils of B or C date to X million years ago and the calculated "family tree" says A was an ancestor of B and C, then A must have evolved more than X million years ago.

It is also possible to estimate how long ago two living clades diverged – i.e. approximately how long ago their last common ancestor must have lived – by assuming that DNA mutations accumulate at a constant rate. These "molecular clocks", however, are fallible, and provide only a very approximate timing: for example, they are not sufficiently precise and reliable for estimating when the groups that feature in the Cambrian explosion first evolved, and estimates produced by different techniques may vary by a factor of two.

History of Life

This wrinkled "elephant skin" texture is a trace fossil of a non-stromatolite microbial mat. The image shows the location, in the Burgsvik beds of Sweden, where the texture was first identified as evidence of a microbial mat.

Earth formed about 4,570 million years ago and, after a collision that formed the Moon about 40 million years later, may have cooled quickly enough to have oceans and an atmosphere about 4,440 million years ago. There is evidence on the Moon of a Late Heavy Bombardment by asteroids from 4,000 to 3,800 million years ago. If, as seems likely, such a bombardment struck Earth at the same time, the first atmosphere and oceans may have been stripped away.

Paleontology traces the evolutionary history of life back to over 3,000 million years ago, possibly as far as 3,800 million years ago. The oldest clear evidence of life on Earth dates to 3,000 million years ago, although there have been reports, often disputed, of fossil bacteria from 3,400 million years ago and of geochemical evidence for the presence of life 3,800 million years ago. Some scientists have proposed that life on Earth was "seeded" from elsewhere, but most research concentrates on various explanations of how life could have arisen independently on Earth.

For about 2,000 million years microbial mats, multi-layered colonies of different bacteria, were the dominant life on Earth. The evolution of oxygenic photosynthesis enabled them to play the major role in the oxygenation of the atmosphere from about 2,400 million years ago. This change in the atmosphere increased their effectiveness as nurseries of evolution. While eukaryotes, cells with complex internal structures, may have been present earlier, their evolution speeded up when they acquired the ability to transform oxygen from a poison to a powerful source of metabolic energy. This innovation may have come from primitive eukaryotes capturing oxygen-powered bacteria as endosymbionts and transforming them into organelles called mitochondria. The earliest evidence of complex eukaryotes with organelles (such as mitochondria) dates from 1,850 million years ago.

Opabinia sparked modern interest in the Cambrian explosion.

Multicellular life is composed only of eukaryotic cells, and the earliest evidence for it is the Francevillian Group Fossils from 2,100 million years ago, although specialisation of cells for different functions first appears between 1,430 million years ago (a possible fungus) and 1,200 million years ago (a probable red alga). Sexual reproduction may be a prerequisite for specialisation of cells, as an asexual multicellular organism might be at risk of being taken over by rogue cells that retain the ability to reproduce.

The earliest known animals are cnidarians from about 580 million years ago, but these are so modern-looking that must be descendants of earlier animals. Early fossils of animals are rare because they had not developed mineralised, easily fossilized hard parts until about 548 million years ago. The earliest modern-looking bilaterian animals appear in the Early Cambrian, along with several "weird wonders" that bear little obvious resemblance to any modern animals. There is a long-running debate about whether this Cambrian explosion was truly a very rapid period of evolutionary experimentation; alternative views are that modern-looking animals began evolving earlier but fossils of their precursors have not yet been found, or that the "weird wonders" are evolutionary "aunts" and "cousins" of modern groups. Vertebrates remained a minor group until the first jawed fish appeared in the Late Ordovician.

At about 13 centimetres (5.1 in) the Early Cretaceous *Yanoconodon* was longer than the average mammal of the time.

The spread of life from water to land required organisms to solve several problems, including protection against drying out and supporting themselves against gravity. The earliest evidence of land plants and land invertebrates date back to about 476 million years ago and 490 million years ago respectively. The lineage that produced land vertebrates evolved later but very rapidly between 370 million years ago and 360 million years ago; recent discoveries have overturned earlier ideas about the history and driving forces behind their evolution. Land plants were so successful that their detritus caused an ecological crisis in the Late Devonian, until the evolution of fungi that could digest dead wood.

Birds are the only surviving dinosaurs.

During the Permian period, synapsids, including the ancestors of mammals, may have dominated land environments, but this ended with the Permian–Triassic extinction event 251 million years ago, which came very close to wiping out all complex life. The extinctions were apparently fairly sudden, at least among vertebrates. During the slow recovery from this catastrophe a previously obscure group, archosaurs, became the most abundant and diverse terrestrial vertebrates. One archosaur group, the dinosaurs, were the dominant land vertebrates for the rest of the Mesozoic, and birds evolved from one group of dinosaurs. During this time mammals' ancestors survived only as small, mainly nocturnal insectivores, which may have accelerated the development of mammalian traits such as endothermy and hair. After the Cretaceous–Paleogene extinction event 66 million years ago killed off all the dinosaurs except the birds, mammals increased rapidly in size and diversity, and some took to the air and the sea.

Fossil evidence indicates that flowering plants appeared and rapidly diversified in the Early Cretaceous between 130 million years ago and 90 million years ago. Their rapid rise to dominance of terrestrial ecosystems is thought to have been propelled by coevolution with pollinating insects. Social insects appeared around the same time and, although they account for only small parts of the insect "family tree", now form over 50% of the total mass of all insects.

Humans evolved from a lineage of upright-walking apes whose earliest fossils date from over 6 million years ago. Although early members of this lineage had chimp-sized brains, about 25% as big as modern humans', there are signs of a steady increase in brain size after about 3 million years ago. There is a long-running debate about whether *modern* humans are descendants of a single small population in Africa, which then migrated all over the world less than 200,000 years ago and replaced previous hominine species, or arose worldwide at the same time as a result of interbreeding.

Mass Extinctions

Life on earth has suffered occasional mass extinctions at least since 542 million years ago. Despite their distrous effects, mass extinctions have sometimes accelerated the evolution of life on earth. When dominance of an ecological niche passes from one group of organisms to another, this is rarely because the new dominant group outcompetes the old, but usually because an extinction event allows new group to outlive the old and move into its niche.

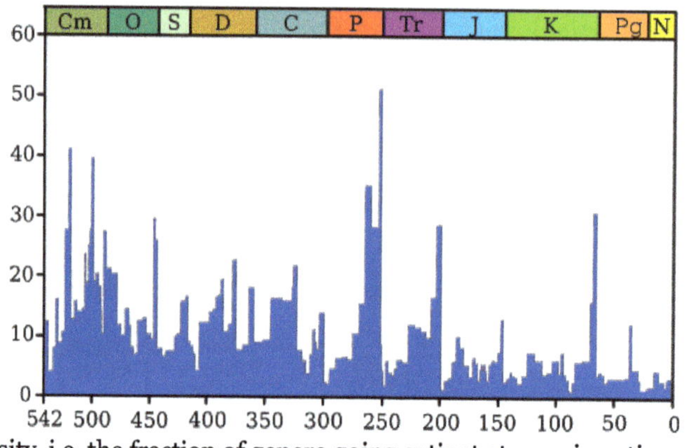

Apparent extinction intensity, i.e. the fraction of genera going extinct at any given time, as reconstructed from the fossil record (graph not meant to include recent epoch of Holocene extinction event)

The fossil record appears to show that the rate of extinction is slowing down, with both the gaps between mass extinctions becoming longer and the average and background rates of extinction decreasing. However, it is not certain whether the actual rate of extinction has altered, since both of these observations could be explained in several ways:

- The oceans may have become more hospitable to life over the last 500 million years and less vulnerable to mass extinctions: dissolved oxygen became more widespread and penetrated to greater depths; the development of life on land reduced the run-off of nutrients and hence the risk of eutrophication and anoxic events; marine ecosystems became more diversified so that food chains were less likely to be disrupted.

- Reasonably complete fossils are very rare, most extinct organisms are represented only by partial fossils, and complete fossils are rarest in the oldest rocks. So paleontologists have mistakenly assigned parts of the same organism to different genera, which were often defined solely to accommodate these finds – the story of Anomalocaris is an example of this. The risk of this mistake is higher for older fossils because these are often unlike parts of any living organism. Many "superfluous" genera are represented by fragments that are not found again, and these "superfluous" genera are interpreted as becoming extinct very quickly.

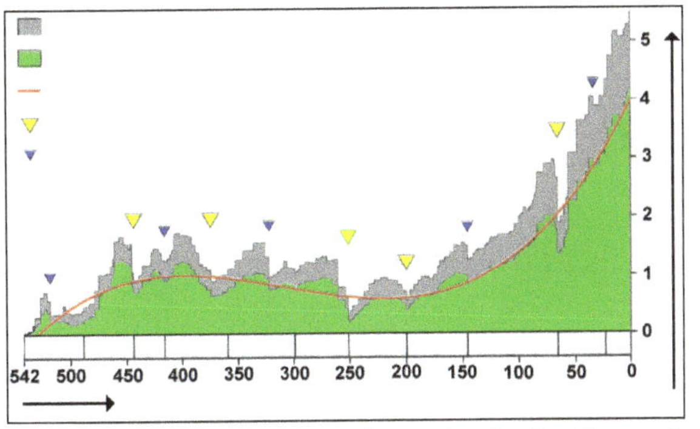

Phanerozoic biodiversity as shown by the fossil record.

Biodiversity in the fossil record, which is:

> "the number of distinct genera alive at any given time; that is, those whose first occurrence predates and whose last occurrence postdates that time"

shows a different trend: a fairly swift rise from 542 to 400 million years ago, a slight decline from 400 to 200 million years ago, in which the devastating Permian–Triassic extinction event is an important factor, and a swift rise from 200 million years ago to the present.

Paleontology

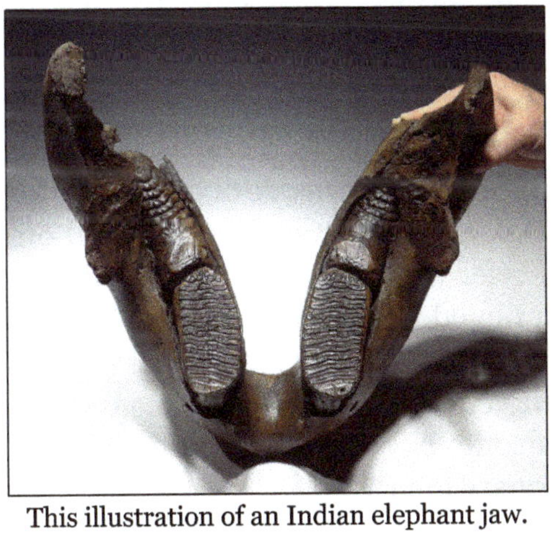

This illustration of an Indian elephant jaw.

Although paleontology became established around 1800, earlier thinkers had noticed aspects of the fossil record. The ancient Greek philosopher Xenophanes (570–480 BC) concluded from fossil sea shells that some areas of land were once under water. During the Middle Ages the Persian naturalist Ibn Sina, known as Avicenna in Europe, discussed fossils and proposed a theory of petrifying fluids on which Albert of Saxony elaborated in the 14th century. The Chinese naturalist Shen Kuo (1031–1095) proposed a theory of climate change based on the presence of petrified bamboo in regions that in his time were too dry for bamboo.

In early modern Europe, the systematic study of fossils emerged as an integral part of the changes in natural philosophy that occurred during the Age of Reason. In the Italian Renaissance, Leonardo Da Vinci made various significant contributions to the field as well depicted numerous fossils. Leonardo's contributions are central to the history of paleontology because he established a line of continuity between the two main branches of paleontology—ichnology and body fossil paleontology. He identified the following:

1. The biogenic nature of ichnofossils, i.e. ichnofossils were structures left by living organisms;
2. The utility of ichnofossils as paleoenvironmental tools—certain ichnofossils show the marine origin of rock strata;
3. The importance of the neoichnological approach—recent traces are a key to understanding ichnofossils;
4. The independence and complementary evidence of ichnofossils and body fossils—ichnofossils are distinct from body fossils, but can be integrated with body fossils to provide paleontological information.

At the end of the 18th century Georges Cuvier's work established comparative anatomy as a scientific discipline and, by proving that some fossil animals resembled no living ones, demonstrated that animals could become extinct, leading to the emergence of paleontology. The expanding knowledge of the fossil record also played an increasing role in the development of geology, particularly stratigraphy.

The first half of the 19th century saw geological and paleontological activity become increasingly well organised with the growth of geologic societies and museums and an increasing number of professional geologists and fossil specialists. Interest increased for reasons that were not purely scientific, as geology and paleontology helped industrialists to find and exploit natural resources such as coal.

This contributed to a rapid increase in knowledge about the history of life on Earth and to progress in the definition of the geologic time scale, largely based on fossil evidence. In 1822 Henri Marie Ducrotay de Blanvill refer to the study of ancient living organisms through fossils. As knowledge of life's history continued to improve, it became increasingly obvious that there had been some kind of successive order to the development of life. This encouraged early evolutionary theories on the transmutation of species. After Charles Darwin published Origin of Species in 1859, much of the focus of paleontology shifted to understanding evolutionary paths, including human evolution, and evolutionary theory.

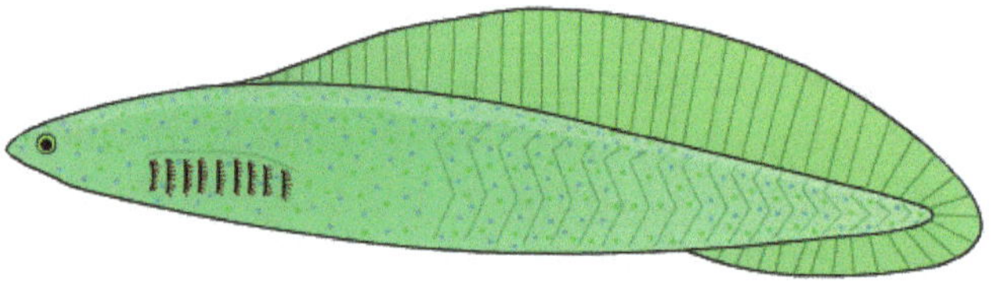

Haikouichthys, from about 518 million years ago in China, may be the earliest known fish.

The last half of the 19th century saw a tremendous expansion in paleontological activity, especially in North America. The trend continued in the 20th century with additional regions of the Earth being opened to systematic fossil collection. Fossils found in China near the end of the 20th century have been particularly important as they have provided new information about the earliest evolution of animals, early fish, dinosaurs and the evolution of birds. The last few decades of the 20th century saw a renewed interest in mass extinctions and their role in the evolution of life on Earth. There was also a renewed interest in the Cambrian explosion that apparently saw the development of the body plans of most animal phyla. The discovery of fossils of the Ediacaran biota and developments in paleobiology extended knowledge about the history of life back far before the Cambrian.

Increasing awareness of Gregor Mendel's pioneering work in genetics led first to the development of population genetics and then in the mid-20th century to the modern evolutionary synthesis, which explains evolution as the outcome of events such as mutations and horizontal gene transfer, which provide genetic variation, with genetic drift and natural selection driving changes in this variation over time. Within the next few years the role and operation of DNA in genetic inheritance were discovered, leading to what is now known as the "Central Dogma" of molecular biology. In the 1960s molecular phylogenetics, the investigation of evolutionary "family trees" by techniques derived from biochemistry, began to make an impact, particularly when it was proposed that the human lineage had diverged from apes much more recently than was generally thought at the time. Although this early study compared proteins from apes and humans, most molecular phylogenetics research is now based on comparisons of RNA and DNA.

Other Branches of Geology

Different branches of geology study one particular part of earth. Since all of the branches are connected, specialists work together to answer complicated questions.

Geochemistry

Geochemistry is the study of the chemical processes which form and shape the Earth. It includes the study of the cycles of matter and energy which transport the Earth's chemical components and the interaction of these cycles with the hydrosphere and the atmosphere.

It is a subfield of inorganic chemistry, which is concerned with the properties of all the elements in the periodic table and their compounds. Inorganic chemistry investigates the characteristics of substances that are not organic, such as nonliving matter and minerals found in the Earth's crust.

Oceanography

Oceanography is the study of the composition and motion of the water column and the processes which are responsible for that motion. The principal oceanographic processes influencing continental shelf waters include waves and tides as well as wind-driven and other oceanic currents. Understanding the oceanography of shelf waters and the influence this has on seabed dynamics, contributes to a wide range of activities such as the following:

- Assessment of offshore petroleum production infrastructure
- Seabed mapping and characterisation for environmental management
- Marine biodiversity surrogacy research
- assessment of renewable energy potential

Paleontology

Paleontologists are interested in fossils and how ancient organisms lived. Paleontology is the study of fossils and what they reveal about the history of our planet. In marine environments, microfossils collected within in layers of sediment cores provide a rich source of information about the environmental history of an area.

Sedimentology

Sedimentology is the study of sediment grains in marine and other deposits, with a focus on physical properties and the processes which form a deposit. Deposition is a geological process where geological material is added to a landform. Key physical properties of interest include:

- The size and shape of sediment grains.
- The degree of sorting of a deposit.
- The composition of grains within a deposit.
- Sedimentary structures.

These properties together provide a record of the mechanisms active during sediment transportation and deposition which allows the interpretation of the environmental conditions that produced a sediment deposit, either in modern settings or in the geological record.

Additional Branches

- Benthic Ecology: Benthic ecology is the study of living things on the seafloor and how they interact with their environment.
- Biostratigraphy: Biostratigraphy is the branch of stratigraphy that uses fossils to establish relative ages of rock and correlate successions of sedimentary rocks within and between depositional basins.
- Geochronology: Geochronology is a discipline of geoscience which measures the age of earth materials and provides the temporal framework in which other geoscience data can be interpreted in the context of Earth history.
- Geophysics: Information relating to various techniques including: airborne electromagnetics, gravity, magnetics, magnetotellurics, radiometrics, rock properties and seismic.
- Marine Geochemistry: Marine geochemistry is the science used to help develop an understanding of the composition of coastal and marine water and sediments.
- Marine Geophysics: Marine geophysics is a scientific discipline which uses the quantitative observation of physical properties to understand the seafloor and sub-seafloor geology.
- Marine Surveying: The survey environment varies from oceanographic studies in the water column to investigating sediment and geochemical processes on the seafloor and imaging the sub-seafloor rocks: Surveys are carried over Australia's entire marine juris-

diction, from coastal estuaries and bays, across the continental shelf and slope, to the deep abyssal plains.

- Spectral Geology: Spectral geology is the measurement and analysis of portions of the electromagnetic spectrum to identify spectrally distinct and physically significant features of different rock types and surface materials, their mineralogy and their alteration signatures.

Plate Tectonics

Plate tectonics, theory dealing with the dynamics of Earth's outer shell—the lithosphere—that revolutionized Earth sciences by providing a uniform context for understanding mountain-building processes, volcanoes, and earthquakes as well as the evolution of Earth's surface and reconstructing its past continents and oceans.

The concept of plate tectonics was formulated in the 1960s. According to the theory, Earth has a rigid outer layer, known as the lithosphere, which is typically about 100 km (60 miles) thick and overlies a plastic (moldable, partially molten) layer called the asthenosphere. The lithosphere is broken up into seven very large continental- and ocean-sized plates, six or seven medium-sized regional plates, and several small ones. These plates move relative to each other, typically at rates of 5 to 10 cm (2 to 4 inches) per year, and interact along their boundaries, where they converge, diverge, or slip past one another. Such interactions are thought to be responsible for most of Earth's seismic and volcanic activity, although earthquakes and volcanoes can occur in plate interiors. Plate motions cause mountains to rise where plates push together, or converge, and continents to fracture and oceans to form where plates pull apart, or diverge. The continents are embedded in the plates and drift passively with them, which over millions of years results in significant changes in Earth's geography.

The theory of plate tectonics is based on a broad synthesis of geologic and geophysical data. It is now almost universally accepted, and its adoption represents a true scientific revolution, analogous in its consequences to quantum mechanics in physics or the discovery of the genetic code in biology. Incorporating the much older idea of continental drift, as well as the concept of seafloor spreading, the theory of plate tectonics has provided an overarching framework in which to describe the past geography of continents and oceans, the processes controlling creation and destruction of landforms, and the evolution of Earth's crust, atmosphere, biosphere, hydrosphere, and climates. During the late 20th and early 21st centuries, it became apparent that plate-tectonic processes profoundly influence the composition of Earth's atmosphere and oceans, serve as a prime cause of long-term climate change, and make significant contributions to the chemical and physical environment in which life evolves.

Principles of Plate Tectonics

In essence, plate-tectonic theory is elegantly simple. Earth's surface layer, 50 to 100 km (30 to 60 miles) thick, is rigid and is composed of a set of large and small plates. Together, these plates constitute the lithosphere, from the Greek lithos, meaning "rock." The lithosphere rests on and slides over an underlying partially molten (and thus weaker but generally denser) layer of plastic partially molten rock known as the asthenosphere, from the Greek asthenos, meaning "weak." Plate movement is possible because the lithosphere-asthenosphere boundary is a zone of detachment. As the lithospheric plates move across Earth's surface, driven by forces as yet not fully understood, they interact along their boundaries, diverging, converging, or slipping past each other. While the interiors of the plates are presumed to remain essentially undeformed, plate boundaries are the sites of many of the principal processes that shape the terrestrial surface, including earthquakes, volcanism, and orogeny (that is, formation of mountain ranges).

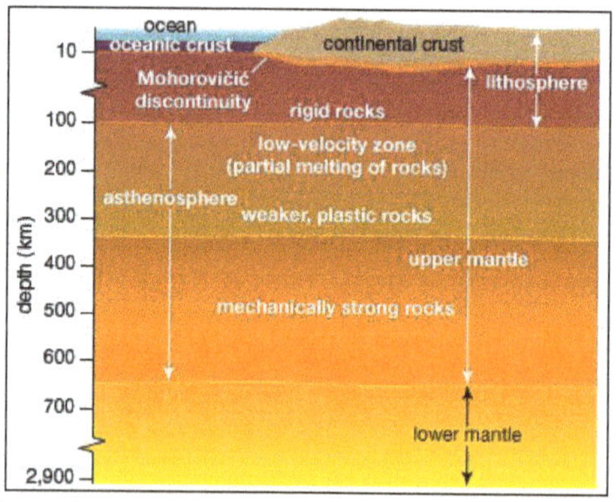

A cross section of Earth's outer layers, from the crust through the lower mantle.

The process of plate tectonics may be driven by convection in Earth's mantle, the pull of heavy old pieces of crust into the mantle, or some combination of both. For a deeper discussion of plate-driving mechanisms.

Earth's Layers

Knowledge of Earth's interior is derived primarily from analysis of the seismic waves that propagate through Earth as a result of earthquakes. Depending on the material they travel through, the waves may either speed up, slow down, bend, or even stop if they cannot penetrate the material they encounter.

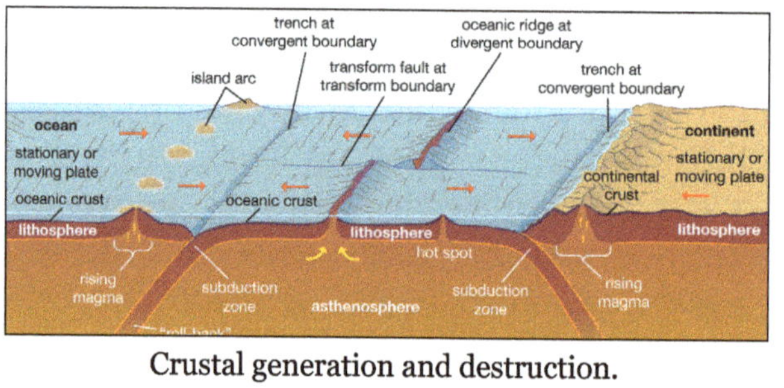

Crustal generation and destruction.

Three-dimensional diagram showing crustal generation and destruction according to the theory of plate tectonics; included are the three kinds of plate boundaries—divergent, convergent (or collision), and strike-slip (or transform).

Collectively, these studies show that Earth can be internally divided into layers on the basis of either gradual or abrupt variations in chemical and physical properties. Chemically, Earth can be divided into three layers. A relatively thin crust, which typically varies from a few kilometres to 40 km (about 25 miles) in thickness, sits on top of the mantle. (In some places, Earth's crust may be up to 70 km [40 miles] thick.) The mantle is much thicker than the crust; it contains 83 percent of Earth's volume and continues to a depth of 2,900 km (1,800 miles). Beneath the mantle is the core, which extends to the centre of Earth, some 6,370 km (nearly 4,000 miles) below the surface. Geologists maintain that the core is made up primarily of metallic iron accompanied by smaller amounts of nickel, cobalt, and lighter elements, such as carbon and sulfur.

There are two types of crust, continental and oceanic, which differ in their composition and thickness. The distribution of these crustal types broadly coincides with the division into continents and ocean basins, although continental shelves, which are submerged, are underlain by continental crust. The continents have a crust that is broadly granitic in composition and, with a density of about 2.7 grams per cubic cm (0.098 pound per cubic inch), is somewhat lighter than oceanic crust, which is basaltic (i.e., richer in iron and magnesium than granite) in composition and has a density of about 2.9 to 3 grams per cubic cm (0.1 to 0.11 pound per cubic inch). Continental crust is typically 40 km (25 miles) thick, while oceanic crust is much thinner, averaging about 6 km (4 miles) in thickness. These crustal rocks both sit on top of the mantle, which is ultramafic in composition (i.e., very rich in magnesium and iron-bearing silicate minerals). The boundary between the crust (continental or oceanic) and the underlying mantle is known as the Mohorovičić discontinuity (also called Moho), which is named for its discoverer, Croatian seismologist Andrija Mohorovičić. The Moho is clearly defined by seismic studies, which detect an acceleration in seismic waves as they pass from the crust into the denser mantle. The boundary between the mantle and the core is also clearly defined by seismic studies, which suggest that the outer part of the core is a liquid.

The effect of the different densities of lithospheric rock can be seen in the different average elevations of continental and oceanic crust. The less-dense continental crust has greater buoyancy, causing it to float much higher in the mantle. Its average elevation above sea level is 840 metres (2,750 feet), while the average depth of oceanic crust is 3,790 metres (12,400 feet). This density difference creates two principal levels of Earth's surface.

The lithosphere itself includes all the crust as well as the upper part of the mantle (i.e., the region directly beneath the Moho), which is also rigid. However, as temperatures increase with depth, the heat causes mantle rocks to lose their rigidity. This process begins at about 100 km (60 miles) below the surface. This change occurs within the mantle and defines the base of the lithosphere and the top of the asthenosphere. This upper portion of the mantle, which is known as the lithospheric mantle, has an average density of about 3.3 grams per cubic cm (0.12 pound per cubic inch). The asthenosphere, which sits directly below the lithospheric mantle, is thought to be slightly denser at 3.4–4.4 grams per cubic cm (0.12–0.16 pound per cubic inch).

In contrast, the rocks in the asthenosphere are weaker, because they are close to their melting temperatures. As a result, seismic waves slow as they enter the asthenosphere. With increasing depth, however, the greater pressure from the weight of the rocks above causes the mantle to become gradually stronger, and seismic waves increase in velocity, a defining characteristic of the lower mantle. The lower mantle is more or less solid, but the region is also very hot, and thus the rocks can flow very slowly (a process known as creep).

During the late 20th and early 21st centuries, scientific understanding of the deep mantle was greatly enhanced by high-resolution seismological studies combined with numerical modeling and laboratory experiments that mimicked conditions near the core-mantle boundary. Collectively, these studies revealed that the deep mantle is highly heterogeneous and that the layer may play a fundamental role in driving Earth's plates.

At a depth of about 2,900 km (1,800 miles), the lower mantle gives way to Earth's outer core, which is made up of a liquid rich in iron and nickel. At a depth of about 5,100 km (3,200 miles), the outer core transitions to the inner core. Although it has a higher temperature than the outer core, the inner core is solid because of the tremendous pressures that exist near Earth's centre. Earth's inner core is divided into the outer-inner core (OIC) and the inner-inner core (IIC), which differ from one another with respect to the polarity of their iron crystals. The polarity of the iron crystals of the OIC is oriented in a north-south direction, whereas that of the IIC is oriented east-west.

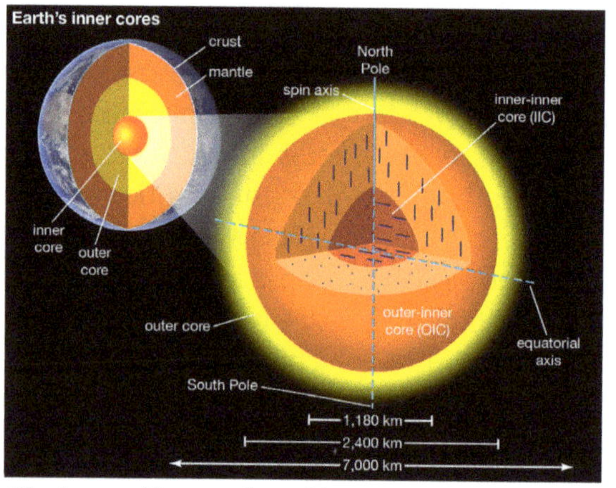

Earth's core The internal layers of Earth's core, including its two inner cores.

Plate Boundaries

Lithospheric plates are much thicker than oceanic or continental crust. Their boundaries do not usually coincide with those between oceans and continents, and their behaviour is only partly influenced by whether they carry oceans, continents, or both. The Pacific Plate, for example, is entirely oceanic, whereas the North American Plate is capped by continental crust in the west (the North American continent) and by oceanic crust in the east and extends under the Atlantic Ocean as far as the Mid-Atlantic Ridge.

In a simplified example of plate motion shown in the figure, movement of plate A to the left relative to plates B and C results in several types of simultaneous interactions along the plate boundaries. At the rear, plates A and B move apart, or diverge, resulting in extension and the formation of a divergent margin. At the front, plates A and B overlap, or converge, resulting in compression and the formation of a convergent margin. Along the sides, the plates slide past one another, a process called shear. As these zones of shear link other plate boundaries to one another, they are called transform faults.

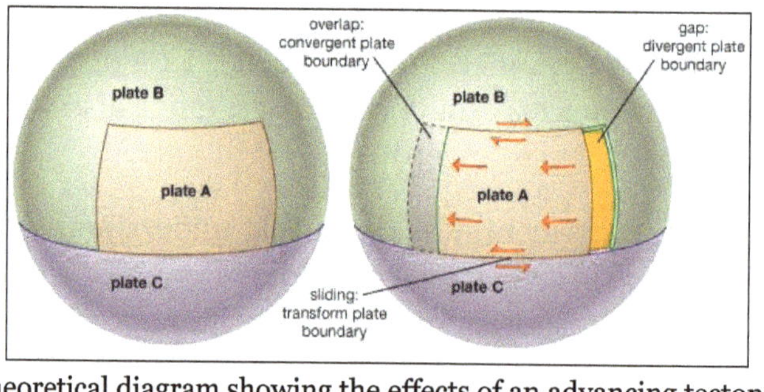

Theoretical diagram showing the effects of an advancing tectonic plate on other adjacent, but stationary, tectonic plates.

At the advancing edge of plate A, the overlap with plate B creates a convergent boundary. In contrast, the gap left behind the trailing edge of plate A forms a divergent boundary with plate B. As plate A slides past portions of both plate B and plate C, transform boundaries develop.

Divergent Margins

As plates move apart at a divergent plate boundary, the release of pressure produces partial melting of the underlying mantle. This molten material, known as magma, is basaltic in composition and is buoyant. As a result, it wells up from below and cools close to the surface to generate new crust. Because new crust is formed, divergent margins are also called constructive margins.

Continental Rifting

Upwelling of magma causes the overlying lithosphere to uplift and stretch. (Whether magmatism [the formation of igneous rock from magma] initiates the rifting or whether rifting decompresses the mantle and initiates magmatism is a matter of significant debate.) If the diverging plates are capped by continental crust, fractures develop that are invaded by the ascending magma, prying the continents farther apart. Settling of the continental blocks creates a rift valley, such as the present-day East African Rift Valley. As the rift continues to widen, the continental crust becomes progressively thinner until separation of the plates is achieved and a new ocean is created. The ascending partial melt cools and crystallizes to form new crust. Because the partial melt is basaltic in composition, the new crust is oceanic, and an ocean ridge develops along the site of the former continental rift. Consequently, diverging plate boundaries, even if they originate within continents, eventually come to lie in ocean basins of their own making.

Rift valley, Thingvellir National Park. The Thingvellir fracture zone at Thingvellir National Park in southwestern Iceland is an example of a rift valley. The Thingvellir fracture lies in the Mid-Atlantic Ridge, which extends through the centre of Iceland.

Seafloor Spreading

As upwelling of magma continues, the plates continue to diverge, a process known as seafloor spreading. Samples collected from the ocean floor show that the age of oceanic crust increases with distance from the spreading centre—important evidence in favour of this process. These age data also allow the rate of seafloor spreading to be determined, and they show that rates vary from about 0.1 cm (0.04 inch) per year to 17 cm (6.7 inches) per year. Seafloor-spreading rates are much more rapid in the Pacific Ocean than in the Atlantic and Indian oceans. At spreading rates of about 15 cm (6 inches) per year, the entire crust beneath the Pacific Ocean (about 15,000 km [9,300 miles] wide) could be produced in 100 million years.

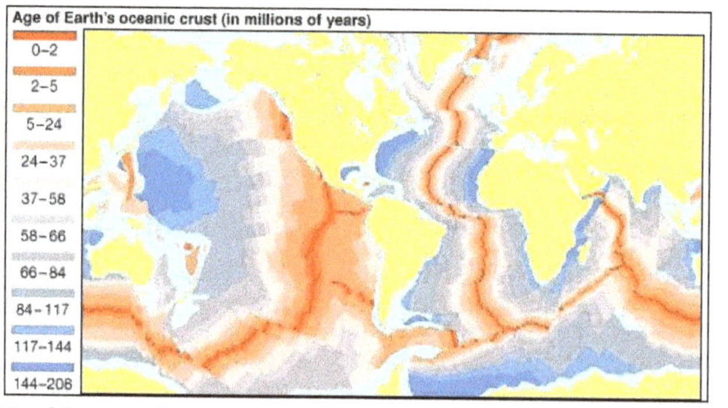

Age of Earth's oceanic crust:Map showing the age of Earth's oceanic crust and the pattern of seafloor spreading at the global scale.

Divergence and creation of oceanic crust are accompanied by much volcanic activity and by many shallow earthquakes as the crust repeatedly rifts, heals, and rifts again. Brittle earthquake-prone rocks occur only in the shallow crust. Deep earthquakes, in contrast, occur less frequently, due to the high heat flow in the mantle rock. These regions of oceanic crust are swollen with heat and so are elevated by 2 to 3 km (1.2 to 1.9 miles) above the surrounding seafloor. The elevated topography results in a feedback scenario in which the resulting gravitational force pushes the crust apart, allowing new magma to well up from below, which in turn sustains the elevated topography. Its summits are typically 1 to 5 km (0.6 to 3.1 miles) below the ocean surface. On a global scale, these ridges form an interconnected system of undersea "mountains" that are about 65,000 km (40,000 miles) in length and are called oceanic ridges.

Convergent Margins

Given that Earth is constant in volume, the continuous formation of Earth's new crust produces an excess that must be balanced by destruction of crust elsewhere. This is accomplished at convergent plate boundaries, also known as destructive plate boundaries, where one plate descends at an angle—that is, is subducted—beneath the other.

Because oceanic crust cools as it ages, it eventually becomes denser than the underlying asthenosphere, and so it has a tendency to subduct, or dive under, adjacent continental plates or younger sections of oceanic crust. The life span of the oceanic crust is prolonged by its rigidity, but eventually this resistance is overcome. Experiments show that the subducted oceanic lithosphere is denser than the surrounding mantle to a depth of at least 600 km (about 400 miles).

The mechanisms responsible for initiating subduction zones are controversial. During the late 20th and early 21st centuries, evidence emerged supporting the notion that subduction zones preferentially initiate along preexisting fractures (such as transform faults) in the oceanic crust. Irrespective of the exact mechanism, the geologic record indicates that the resistance to subduction is overcome eventually.

Where two oceanic plates meet, the older, denser plate is preferentially subducted beneath the younger, warmer one. Where one of the plate margins is oceanic and the other is continental, the greater buoyancy of continental crust prevents it from sinking, and the oceanic plate is preferentially subducted. Continents are preferentially preserved in this manner relative to oceanic crust, which is continuously recycled into the mantle. This explains why ocean floor rocks are generally less than 200 million years old whereas the oldest continental rocks are more than 4 billion years old. Before the middle of the 20th century, most geoscientists maintained that continental crust was too buoyant to be subducted. However, it later became clear that slivers of continental crust adjacent to the deep-sea trench, as well as sediments deposited in the trench, may be dragged down the subduction zone. The recycling of this material is detected in the chemistry of volcanoes that erupt above the subduction zone.

Two plates carrying continental crust collide when the oceanic lithosphere between them has been eliminated. Eventually, subduction ceases and towering mountain ranges, such as the Himalayas, are created.

Because the plates form an integrated system, it is not necessary that new crust formed at any given divergent boundary be completely compensated at the nearest subduction zone, as long as the total amount of crust generated equals that destroyed.

Subduction Zones

The subduction process involves the descent into the mantle of a slab of cold hydrated oceanic lithosphere about 100 km (60 miles) thick that carries a relatively thin cap of oceanic sediments. The path of descent is defined by numerous earthquakes along a plane that is typically inclined between 30° and 60° into the mantle and is called the Wadati-Benioff zone, for Japanese seismologist Kiyoo Wadati and American seismologist Hugo Benioff, who pioneered its study. Between 10 and 20 percent of the subduction zones that dominate the circum-Pacific ocean basin are subhorizontal (that is, they subduct at angles between 0° and 20°). The factors that govern the dip of the subduction zone are not fully understood, but they probably include the age and thickness of the subducting oceanic lithosphere and the rate of plate convergence.

Most, but not all, earthquakes in this planar dipping zone result from compression, and the seismic activity extends 300 to 700 km (200 to 400 miles) below the surface, implying that the subducted crust retains some rigidity to this depth. At greater depths the subducted plate is partially recycled into the mantle.

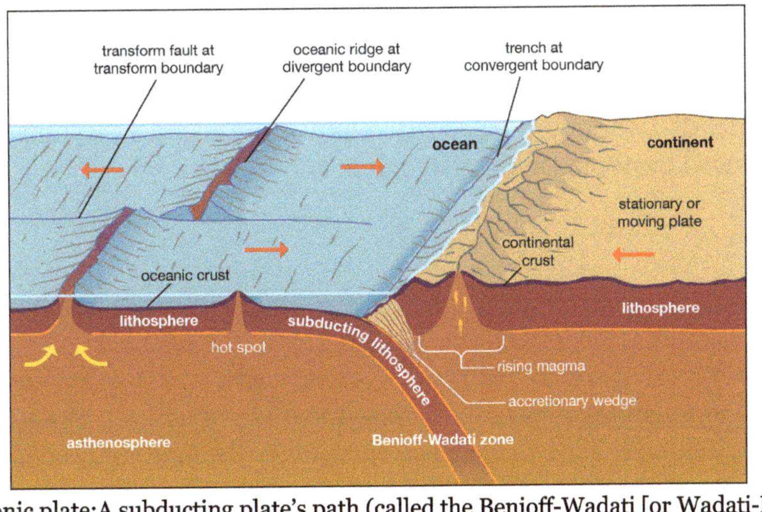

Subducting tectonic plate:A subducting plate's path (called the Benioff-Wadati [or Wadati-Benioff] zone) is defined by numerous earthquakes along a plane that is typically inclined between 30° and 60° into the mantle.

The site of subduction is marked by a deep trench, between 5 and 11 km (3 and 7 miles) deep, that is produced by frictional drag between the plates as the descending plate bends before it subducts. The overriding plate scrapes sediments and elevated portions of ocean floor off the upper crust of the lower plate, creating a zone of highly deformed rocks within the trench that becomes attached, or accreted, to the overriding plate. This chaotic mixture is known as an accretionary wedge.

The rocks in the subduction zone experience high pressures but relatively low temperatures, an effect of the descent of the cold oceanic slab. Under these conditions the rocks recrystallize, or metamorphose, to form a suite of rocks known as blueschists, named for the diagnostic blue mineral called glaucophane, which is stable only at the high pressures and low temperatures found in subduction zones. At deeper levels in the subduction zone (that is, greater than 30–35 km [about 19–22 miles]), eclogites, which consist of high-pressure minerals such as red garnet (pyrope) and omphacite (pyroxene), form. The formation of eclogite from blueschist is accompanied by a significant increase in density and has been recognized as an important additional factor that facilitates the subduction process.

Island Arcs

When the downward-moving slab reaches a depth of about 100 km (60 miles), it gets sufficiently warm to drive off its most volatile components, thereby stimulating partial melting of mantle in the plate above the subduction zone (known as the mantle wedge). Melting in the mantle wedge produces magma, which is predominantly basaltic in composition. This magma rises to the surface and gives birth to a line of volcanoes in the overriding plate, known as a volcanic arc, typically a few hundred kilometres behind the oceanic trench. The distance between the trench and the arc, known as the arc-trench gap, depends on the angle of subduction. Steeper subduction zones have relatively narrow arc-trench gaps. A basin may form within this region, known as a fore-arc basin, and may be filled with sediments derived from the volcanic arc or with remains of oceanic crust.

If both plates are oceanic, as in the western Pacific Ocean, the volcanoes form a curved line of islands, known as an island arc, that is parallel to the trench, as in the case of the Mariana Islands and the adjacent Mariana Trench. If one plate is continental, the volcanoes form inland, as they do in the Andes of western South America. Though the process of magma generation is similar, the ascending magma may change its composition as it rises through the thick lid of continental crust, or it may provide sufficient heat to melt the crust. In either case, the composition of the volcanic mountains formed tends to be more silicon-rich and iron- and magnesium-poor relative to the volcanic rocks produced by ocean-ocean convergence.

Back-Arc Basins

Where both converging plates are oceanic, the margin of the older oceanic crust will be subducted because older oceanic crust is colder and therefore more dense. As the dense slab collapses into the asthenosphere, however, it also may "roll back" oceanward and cause extension in the overlying plate. This results in a process known as back-arc spreading, in which a basin opens up behind the island arc. The crust behind the arc becomes progressively thinner, and the decompression of the underlying mantle causes the crust to melt, initiating seafloor-spreading processes, such as melting and the production of basalt; these processes are similar to those that occur at ocean ridges. The geochemistry of the basalts produced at back-arc basins superficially resembles that of basalts produced at ocean ridges, but subtle trace element analyses can detect the influence of a nearby subducted slab.

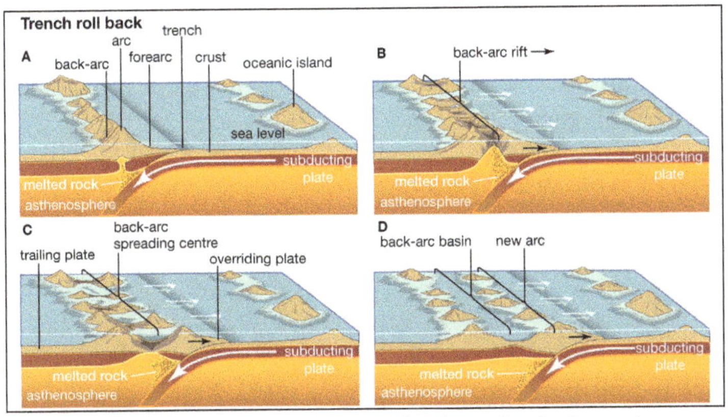

The trench "roll back" process of back-arc basin formation.

This style of subduction predominates in the western Pacific Ocean, in which a number of back-arc basins separate several island arcs from Asia. Examples include the Mariana Islands, the Kuril Islands, and the main islands of Japan. However, if the rate of convergence increases or if anomalously thick oceanic crust (possibly caused by rising mantle plume activity) is conveyed into the subduction zone, the slab may flatten. Such flattening causes the back-arc basin to close, resulting in deformation, metamorphism, and even melting of the strata deposited in the basin.

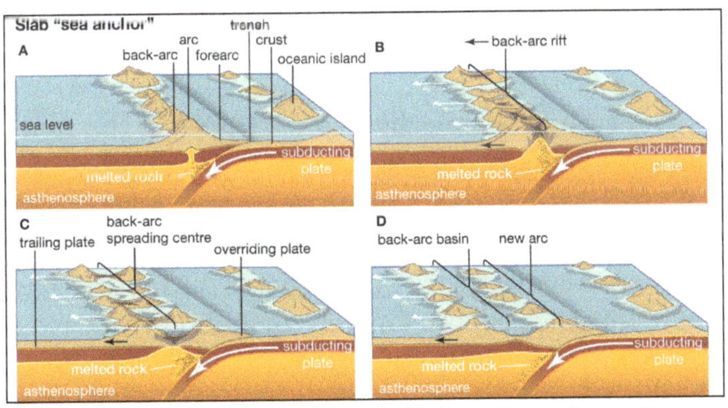

The slab "sea anchor" process of back-arc basin formation.

Mountain Building

If the rate of subduction in an ocean basin exceeds the rate at which the crust is formed at oceanic ridges, a convergent margin forms as the ocean initially contracts. This process can lead to collision between the approaching continents, which eventually terminates subduction. Mountain building can occur in a number of ways at a convergent margin: mountains may rise as a consequence of the subduction process itself, by the accretion of small crustal fragments (which, along with linear island chains and oceanic ridges, are known as terranes), or by the collision of two large continents.

Many mountain belts were developed by a combination of these processes. For example, the Cordilleran mountain belt of North America—which includes the Rocky Mountains as well as the Cascades, the Sierra Nevada, and other mountain ranges near the Pacific coast—developed by a combination of subduction and terrane accretion. As continental collisions are usually preceded by a long history of subduction and terrane accretion, many mountain belts record all three processes. Over the past 70 million years the subduction of the Neo-Tethys Sea, a wedge-shaped body of water that was located between Gondwana and Laurasia, led to the accretion of terranes along the margins of Laurasia, followed by continental collisions beginning about 30 million years ago between Africa and Europe and between India and Asia. These collisions culminated in the formation of the Alps and the Himalayas.

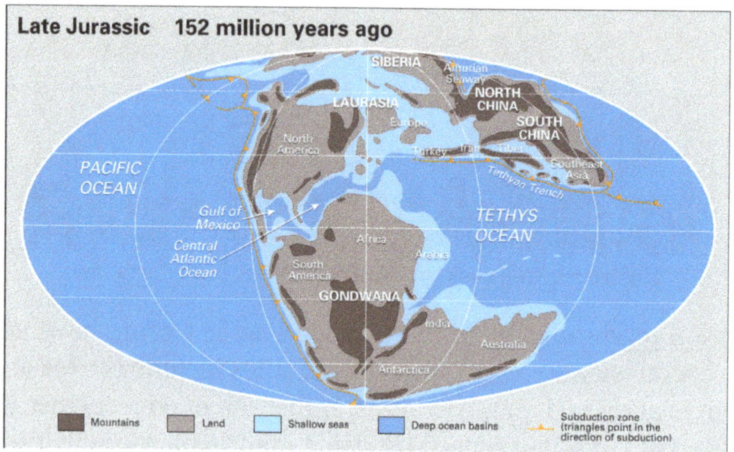

Jurassic paleogeography Distribution of landmasses, mountainous regions, shallow seas, and deep ocean basins during the late Jurassic Period. Included in the paleogeographic reconstruction are the locations of the interval's subduction zones.

Mountains by Subduction

Mountain building by subduction is classically demonstrated in the Andes Mountains of South America. Subduction results in voluminous magmatism in the mantle and crust overlying the subduction zone, and, therefore, the rocks in this region are warm and weak. Although subduction is a long-term process, the uplift that results in mountains tends to occur in discrete episodes and may reflect intervals of stronger plate convergence that squeezes the thermally weakened crust upward. For example, rapid uplift of the Andes approximately 25 million years ago is evidenced by a reversal in the flow of the Amazon River from its ancestral path toward the Pacific Ocean to its modern path, which empties into the Atlantic Ocean.

In addition, models have indicated that the episodic opening and closing of back-arc basins have been the major factors in mountain-building processes, which have influenced the plate-tectonic evolution of the western Pacific for at least the past 500 million years.

Mountains by Terrane Accretion

As the ocean contracts by subduction, elevated regions within the ocean basin—terranes—are transported toward the subduction zone, where they are scraped off the descending plate and added—accreted—to the continental margin. Since the late Devonian and early Carboniferous periods, some 360 million years ago, subduction beneath the western margin of North America has resulted in several collisions with terranes. The piecemeal addition of these accreted terranes has added an average of 600 km (400 miles) in width along the western margin of the North American continent, and the collisions have resulted in important pulses of mountain building.

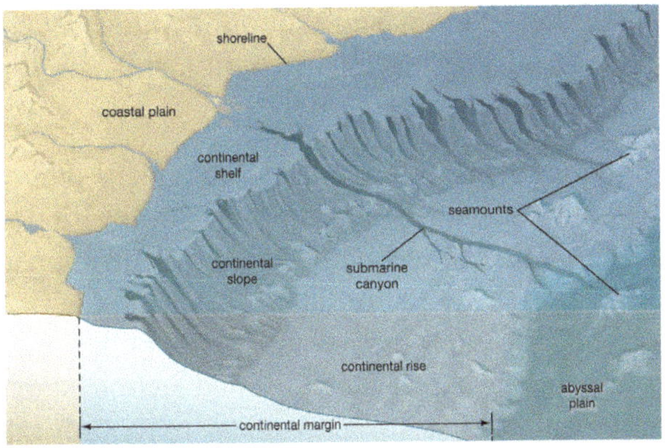

Continental margin: The broad, gentle pitch of the continental shelf gives way to the relatively steep continental slope. The more gradual transition to the abyssal plain is a sediment-filled region called the continental rise. The continental shelf, slope, and rise are collectively called the continental margin.

During these accretionary events, small sections of the oceanic crust may break away from the subducting slab as it descends. Instead of being subducted, these slices are thrust over the overriding plate and are said to be obducted. Where this occurs, rare slices of ocean crust, known as ophiolites, are preserved on land. They provide a valuable natural laboratory for studying the composition and character of the oceanic crust and the mechanisms of their emplacement and preservation on land. A classic example is the Coast Range ophiolite of California, which is one of the most extensive ophiolite terranes in North America. These ophiolite deposits run from the Klamath Mountains in northern California southward to the Diablo Range in central California. This oceanic crust likely formed during the middle of the Jurassic Period, roughly 170 million years ago, in an extensional regime within either a back-arc or a forearc basin. In the late Mesozoic, it was accreted to the western North American continental margin.

Because preservation of oceanic crust is rare, the recognition of ophiolite complexes is very important in tectonic analyses. Until the mid-1980s, ophiolites were thought to represent vestiges of the main oceanic tract, but geochemical analyses have clearly indicated that most ophiolites form near volcanic arcs, such as in back-arc basins characterized by subduction roll-back (the collapse of the subducting plate that causes the extension of the overlying plate). The recognition of ophiolite complexes is very important in tectonic analysis, because they provide insights into the generation of magmatism in oceanic domains, as well as their complex relationships with subduction processes.

Mountains by Continental Collision

Continental collision involves the forced convergence of two buoyant plate margins that results in neither continent being subducted to any appreciable extent. A complex sequence of events ensues that compels one continent to override the other. These processes result in crustal thickening and intense deformation that forces the crust skyward to form huge mountains with crustal roots that extend as deep as 80 km (about 50 miles) relative to Earth's surface, in accordance with the principles of isostasy.

The subducted slab still has a tendency to sink and may become detached and founder (submerge) into the mantle. The crustal root undergoes metamorphic reactions that result in a significant increase in density and may cause the root to also founder into the mantle. Both processes result in a significant injection of heat from the compensatory upwelling of asthenosphere, which is an important contribution to the rise of the mountains.

Continental collisions produce lofty landlocked mountain ranges such as the Himalayas. Much later, after these ranges have been largely leveled by erosion, it is possible that the original contact, or suture, may be exposed.

The balance between creation and destruction on a global scale is demonstrated by the expansion of the Atlantic Ocean by seafloor spreading over the past 200 million years, compensated by the contraction of the Pacific Ocean, and the consumption of an entire ocean between India and Asia (the Tethys Sea). The northward migration of India led to collision with Asia some 40 million years ago. Since that time India has advanced a further 2,000 km (1,250 miles) beneath Asia, pushing up the Himalayas and forming the Plateau of Tibet. Pinned against stable Siberia, China and Indochina were pushed sideways, resulting in strong seismic activity thousands of kilometres from the site of the continental collision.

Transform Faults

Along the third type of plate boundary, two plates move laterally and pass each other along giant fractures in Earth's crust. Transform faults are so named because they are linked to other types of plate boundaries. The majority of transform faults link the offset segments of oceanic ridges. However, transform faults also occur between plate margins with continental crust—for example, the San Andreas Fault in California and the North Anatolian fault system in Turkey. These boundaries are conservative because plate interaction occurs without creating or destroying crust. Because the only motion along these faults is the sliding of plates past each other, the horizontal direction along the fault surface must parallel the direction of plate motion. The fault surfaces are rarely smooth, and pressure may build up when the plates on either side temporarily lock. This buildup of stress may be suddenly released in the form of an earthquake.

Section of the San Andreas Fault in the Carrizo Plain.

Many transform faults in the Atlantic Ocean are the continuation of major faults in adjacent continents, which suggests that the orientation of these faults might be inherited from preexisting weaknesses in continental crust during the earliest stages of the development of oceanic crust. On the other hand, transform faults may themselves be reactivated, and recent geodynamic models suggest that they are favourable environments for the initiation of subduction zones.

Hotspots

Although most of Earth's volcanic activity is concentrated along or adjacent to plate boundaries, there are some important exceptions in which this activity occurs within plates. Linear chains of islands, thousands of kilometres in length, that occur far from plate boundaries are the most notable examples. These island chains record a typical sequence of decreasing elevation along the chain, from volcanic island to fringing reef to atoll and finally to submerged seamount. An active volcano usually exists at one end of an island chain, with progressively older extinct volcanoes occurring along the rest of the chain. Canadian geophysicist J. Tuzo Wilson and American geophysicist W. Jason Morgan explained such topographic features as the result of hotspots.

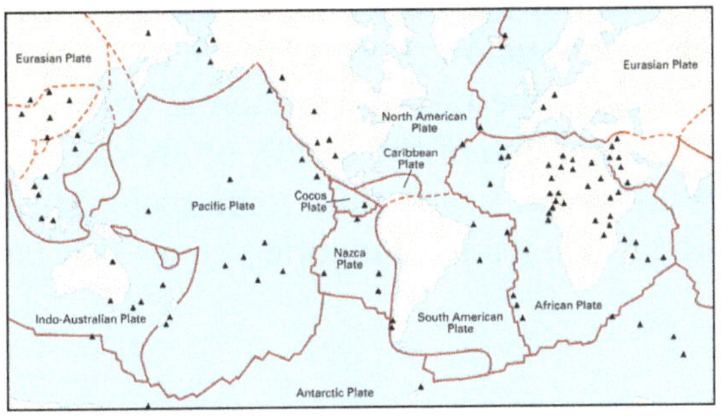

The principal tectonic plates that make up Earth's lithosphere. Also located are several dozen hot spots where plumes of hot mantle material are upwelling beneath the plates.

The number of these hotspots is uncertain (estimates range from 20 to 120), but most occur within a plate rather than at a plate boundary. Hotspots are thought to be the surface expression of giant plumes of heat, termed mantle plumes, that ascend from deep within the mantle, possibly from the core-mantle boundary, some 2,900 km (1,800 miles) below the surface. These plumes are thought to be stationary relative to the lithospheric plates that move over them. A volcano builds upon the surface of a plate directly above the plume. As the plate moves on, however, the volcano is separated from its underlying magma source and becomes extinct. Extinct volcanoes are eroded as they cool and subside to form fringing reefs and atolls, and eventually they sink below the surface of the sea to form a seamount. At the same time, a new active volcano forms directly above the mantle plume.

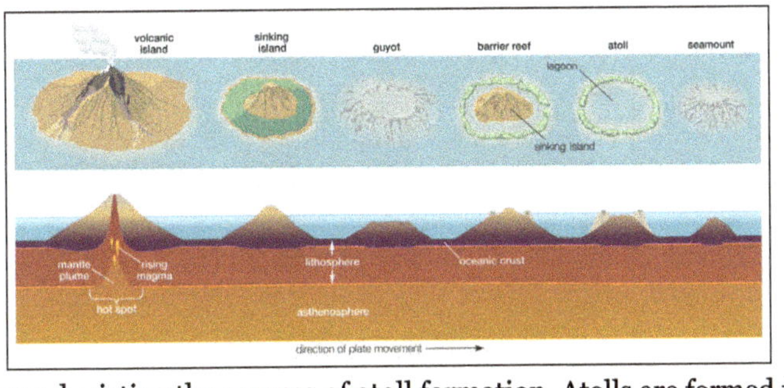

Diagram depicting the process of atoll formation. Atolls are formed from the remnant parts of sinking volcanic islands.

The best example of this process is preserved in the Hawaiian-Emperor seamount chain. The plume is presently situated beneath Hawaii, and a linear chain of islands, atolls, and seamounts extends 3,500 km (2,200 miles) northwest to Midway and a further 2,500 km (1,500 miles) north-north-west to the Aleutian Trench. The age at which volcanism became extinct along this chain gets progressively older with increasing distance from Hawaii—critical evidence that supports this theory. Hotspot volcanism is not restricted to the ocean basins; it also occurs within continents, as in the case of Yellowstone National Park in western North America.

Measurements suggest that hotspots may move relative to one another, a situation not predicted by the classical model, which describes the movement of lithospheric plates over stationary mantle plumes. This has led to challenges to this classic model. Furthermore, the relationship between hotspots and plumes is hotly debated. Proponents of the classical model maintain that these discrepancies are due to the effects of mantle circulation as the plumes ascend, a process called the mantle wind. Data from alternative models suggest that many plumes are not deep-rooted. Instead, they provide evidence that many mantle plumes occur as linear chains that inject magma into fractures, result from relatively shallow processes such as the localized presence of water-rich mantle, stem from the insulating properties of continental crust (which leads to the buildup of trapped mantle heat and decompression of the crust), or are due to instabilities in the interface between continental and oceanic crust. In addition, some geologists note that many geologic processes that others attribute to the behaviour of mantle plumes may be explained by other forces.

Plate Motion

Euler's Contributions

In the 18th century, Swiss mathematician Leonhard Euler showed that the movement of a rigid body across the surface of a sphere can be described as a rotation (or turning) around an axis that goes through the centre of the sphere, known as the axis of rotation. The location of this axis bears no relationship to Earth's spin axis. The point of emergence of the axis through the surface of the sphere is known as the pole of rotation. This theorem of spherical geometry provides an elegant way to define the motion of the lithospheric plates across Earth's surface. Therefore, the relative motion of two rigid plates may be described as rotations around a common axis, known as the axis of spreading. Application of the theorem requires that the plates not be internally deformed—a requirement not absolutely adhered to but one that appears to be a reasonable approximation of what actually happens. Application of this theorem permits the mathematical reconstruction of past plate configurations.

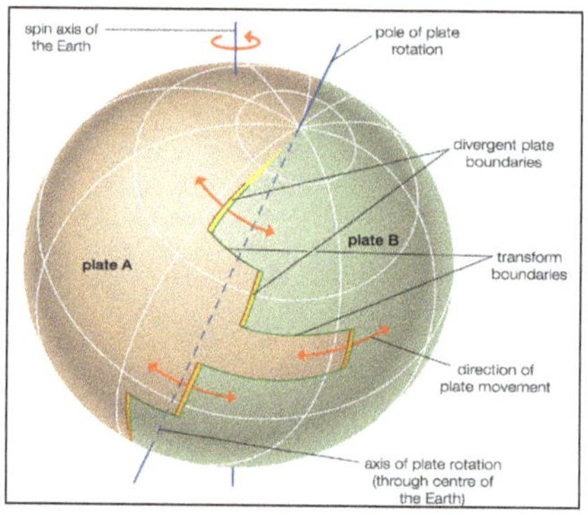

Theoretical depiction of the movement of tectonic plates across Earth's surface.

Movement on a sphere of two plates, A and B, can be described as a rotation around a common pole. Circles around that pole correspond to the orientation of transform faults (that is, single lines in the horizontal that connect to divergent plate boundaries, marked by double lines, in the vertical).

Because all plates form a closed system, all movements can be defined by dealing with them two at a time. The joint pole of rotation of two plates can be determined from their transform boundaries, which are by definition parallel to the direction of motion. Thus, the plates move along transform faults, whose trace defines circles of latitude perpendicular to the axis of spreading, and so form small circles around the pole of rotation. A geometric necessity of this theorem—that lines perpendicular to the transform faults converge on the pole of rotation—is confirmed by measurements. According to this theorem, the rate of plate motion should be slowest near the pole of rotation and increase progressively to a maximum rate along fractures with a 90° angle to it. This relationship is also confirmed by accurate measurements of seafloor-spreading rates.

Past Plate Movements

Plate tectonics involves the movements of Earth's lithospheric plates relative to one another over the planet's weak asthenosphere. This activity changes the positions of all plates with respect to Earth's spin axis and the Equator. To determine the true geographic positions of the plates in the past, investigators have to define their motions, not only relative to each other but also relative to this independent frame of reference. Hotspots, as classically interpreted, provide an example of such a reference frame, assuming they are the sources of plumes that originate within the deep mantle and have relatively fixed positions over time. If this assumption is valid, the motion of the lithosphere above these plumes can be deduced. The hotspot island chains serve this purpose, their trends providing the direction of motion of a plate. The speed of the plate can be inferred from the increase in age of the volcanoes along the chain relative to the distance between the islands.

Earth scientists are able to accurately reconstruct the positions and movements of plates for the past 150 million to 200 million years because they have the oceanic crust record to provide them with plate speeds and direction of movement. However, since older oceanic crust is continuously consumed to make room for new crust, this kind of evidence is not available for earlier intervals of geologic time, making it necessary for investigators to turn to other, less-precise techniques.

Development of Tectonic Theory

Precursors

The outlines of the continents flanking the Atlantic Ocean are so similar that their correspondence was apparent as soon as accurate maps became available. The earliest references to this similarity were made in 1596 by Flemish cartographer Abraham Ortelius and later in 1620 by the English philosopher Francis Bacon, in his book Novum Organum, and by French naturalist Georges-Louis Leclerc, count de Buffon, a century later. Toward the end of the 18th century, Alexander von Humboldt, a German naturalist, suggested that the lands bordering the Atlantic Ocean had once been joined.

In 1858 French geographer Antonio Snider-Pellegrini proposed that identical fossil plants in North American and European coal deposits could be explained if the two continents had formerly been connected. He suggested that the biblical Flood was due to the fragmentation of this continent, which was torn apart to restore the balance of a lopsided Earth. In the late 19th century the Austrian geologist Eduard Suess proposed that large ancient continents had been composed of several of the present-day smaller ones. According to this hypothesis, portions of a single enormous southern continent—designated Gondwana (or Gondwanaland)—foundered to create the Atlantic and Indian oceans. Such sunken lands, along with vanished land bridges, were frequently invoked in the late 1800s to explain sediment sources apparently present in the ocean and to account for floral and faunal connections between continents. These explanations remained popular until the 1950s and stimulated belief in the ancient submerged continent of Atlantis.

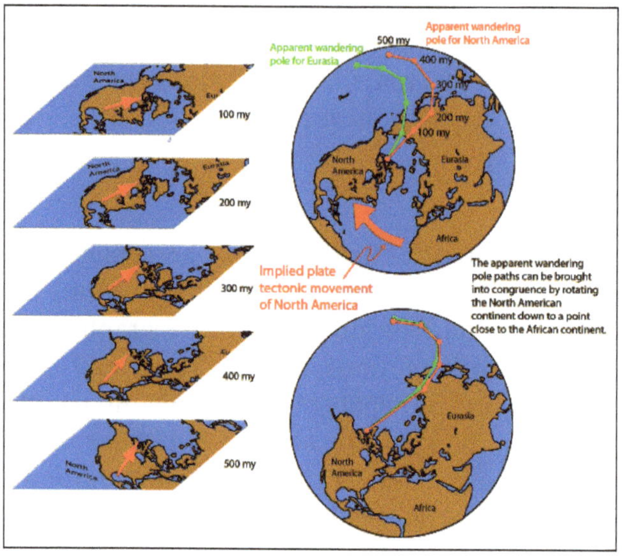

Paleocontinent Infographic showing evidence of submerged continents that formed and broke up during Earth's geologic history.

In 1908 American geologist Frank B. Taylor postulated that the arcuate (bow-shaped) mountain belts of Asia and Europe resulted from the creep of the continents toward the Equator. His analysis of tectonic features foreshadowed in many ways modern thought regarding plate collisions.

Alfred Wegener and the Concept of Continental Drift

In 1912 German meteorologist Alfred Wegener, impressed by the similarity of the geography of the Atlantic coastlines, explicitly presented the concept of continental drift. Though plate tectonics is by no means synonymous with continental drift, the term encompasses this idea and derives much of its impact from it.

Wegener, Alfred Alfred Wegener, German geophysicist and meteorologist.

Wegener came to consider the existence of a single supercontinent from about 350 million to 245 million years ago, during the late Paleozoic Era and early Mesozoic Era, and named it Pangea, meaning "all lands." He searched the geologic and paleontological literature for evidence supporting the continuity of geologic features across the Indian and Atlantic oceans during that time period, which he assumed had formed during the Mesozoic Era (about 252 million to 66 million years ago).

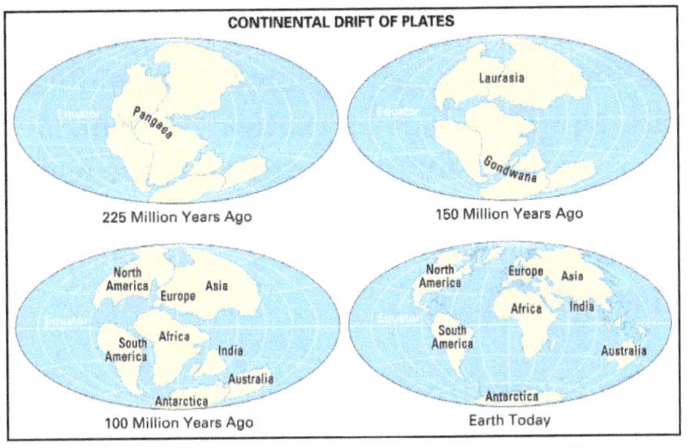

The location of Earth's continents at various times between 225 million years ago and the present.

Isostasy

Wegener pointed out that the concept of isostasy (the ideal theoretical balance of all large portions of Earth's lithosphere as though they were floating on the denser underlying layer) rendered the existence of large sunken continental blocks, as envisaged by Suess, geophysically impossible. He concluded that if the continents had been once joined together, the consequence would have been drift of their fragments and not their foundering. The assumption of a former single continent could be tested geologically, and Wegener displayed a large array of data that supported his hypothesis, ranging from the continuity of fold belts across oceans, the presence of identical rocks and fossils on continents now separated by oceans, and the paleobiogeographic and paleoclimatological record that indicated otherwise unaccountable shifts in Earth's major climate belts. He further argued that if continents could move up and down in the mantle as a result of buoyancy changes produced by erosion or deposition, they should be able to move horizontally as well.

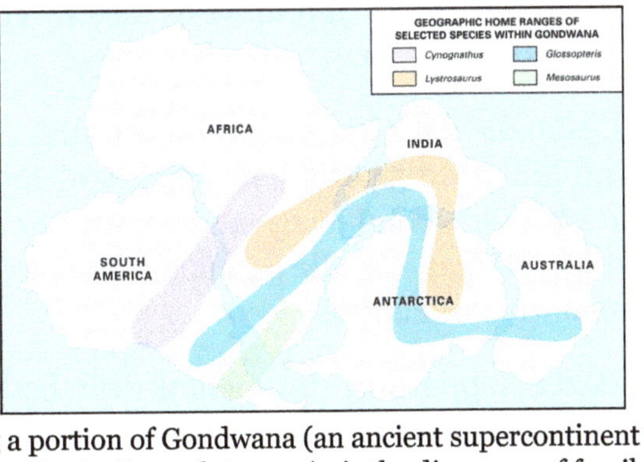

In this map depicting a portion of Gondwana (an ancient supercontinent that once contained South America, Africa, Australia, India, and Antarctica), the discovery of fossil plants and animals whose geographic home ranges cut across the greater landmass is supporting evidence for continental drift.

Driving Forces

The main stumbling block to the acceptance of Wegener's hypothesis was the driving forces he proposed. Wegener described the drift of continents as a flight from the poles due to Earth's equatorial bulge. Although these forces do exist, Wegener's nemesis, British geophysicist Sir Harold

Jeffreys, demonstrated that these forces are much too weak for the task. Another mechanism proposed by Wegener, tidal forces on Earth's crust produced by gravitational pull of the Moon, were also shown to be entirely inadequate.

Wegener's proposition was attentively received by many European geologists, and in England Arthur Holmes pointed out that the lack of a driving force was insufficient grounds for rejecting the entire concept. In 1929 Holmes proposed an alternative mechanism—convection of the mantle—which remains today a serious candidate for the force driving the plates. Wegener's ideas also were well received by geologists in the Southern Hemisphere. One of them, the South African Alexander Du Toit, remained an ardent believer. After Wegener's death, Du Toit continued to amass further evidence in support of continental drift.

Evidence Supporting the Hypothesis

The strikingly similar Paleozoic sedimentary sequences on all southern continents and also in India are an example of evidence that supports continental drift. This diagnostic sequence consists of glacial deposits called tillites, followed by sandstones and finally coal measures, typical of warm moist climates. An attempt to explain this sequence in a world of fixed continents presents insurmountable problems. Placed on a reconstruction of Gondwana, however, the tillites mark two ice ages that occurred during the drift of this continent across the South Pole from its initial position north of Libya about 500 million years ago and its final departure from southern Australia 250 million years later. About this time, Gondwana collided with Laurentia (the precursor to the North American continent), which was one of the major collisional events that produced Pangea.

Both ice ages resulted in glacial deposits—in the southern Sahara during the Silurian Period (443.8 million to 419.2 million years ago) and in southern South America, South Africa, India, and Australia from 382.7 million to 251.9 million years ago, spanning the latter part of the Devonian, as well as the Carboniferous and the Permian. At each location, the tillites were subsequently covered by desert sands of the subtropics and these in turn by coal measures, indicating that the region had arrived near the paleoequator.

During the 1950s and 1960s, isotopic dating of rocks showed that the crystalline massifs of Precambrian age (from about 4.6 billion to 541 million years ago) found on opposite sides of the South Atlantic did indeed closely correspond in age and composition, as Wegener had surmised. It is now evident that they originated as a single assemblage of Precambrian continental nuclei later torn apart by the fragmentation of Pangea.

By the 1960s, although evidence supporting continental drift had strengthened substantially, many scientists were claiming that the shape of the coastlines should be more sensitive to coastal erosion and changes in sea level and are unlikely to maintain their shape over hundreds of millions of years. Therefore, they argued, the supposed fit of the continents flanking the Atlantic Ocean is fortuitous. In 1964, however, those arguments were laid to rest. A computer analysis by Sir Edward Bullard showed an impressive fit of these continents at the 1,000-metre (3,300-foot) depth contour, which strongly supported the notion that Africa and South America were once joined together. A match at this depth is highly significant and is a better approximation of the edge of the continents than the present shoreline. With this reconstruction of the continents, the structures and stratigraphic sequences of Paleozoic mountain ranges in eastern North America and northwestern Europe can be matched in detail in the manner envisaged by Wegener.

Disbelief and Opposition

Sir Harold Jeffreys was one of the strongest opponents of Wegener's hypothesis. He believed that continental drift is impossible because the strength of the underlying mantle should be far greater than any conceivable driving force. In North America, opposition to Wegener's ideas was most vigorous and very nearly unanimous. Wegener was attacked from virtually every possible vantage point—his paleontological evidence attributed to land bridges, the similarity of strata (layers of rock) on both sides of the Atlantic called into question, and the fit of Atlantic shores declared inaccurate. This criticism is illustrated by reports from a symposium on continental drift organized in 1928 by the American Association of Petroleum Geologists. The mood that prevailed at the gathering was expressed by an unnamed attendant quoted with sympathy by the great American geologist Thomas C. Chamberlin: "If we are to believe Wegener's hypothesis, we must forget everything which has been learned in the last 70 years and start all over again." The same reluctance to start anew was again displayed some 40 years later by the same organization when its publications provided the principal forum for the opposition to plate tectonics.

Renewed Interest in Continental Drift

Paleomagnetism, Polar Wandering, and Continental Drift

Ironically, the final vindication of Wegener's hypothesis came from the field of geophysics, the subject used by Jeffreys to discredit the original concept. The ancient Greeks realized that some rocks are strongly magnetized, and the Chinese invented the magnetic compass in the 13th century. In the 19th century geologists recognized that many rocks preserve the imprint of Earth's magnetic field as it was at the time of their formation. The study and measurement of Earth's ancient magnetic field and the remanent magnetism in rocks is called paleomagnetism. Iron-rich volcanic rocks, such as basalt, contain minerals that are good recorders of remanent magnetism, and some sediments also align their magnetic particles with Earth's field at the time of deposition. These minerals behave like fossil compasses that indicate, like any magnet suspended in Earth's field, the direction to the magnetic pole and the latitude of their origin at the time the minerals were crystallized or deposited.

During the 1950s, paleomagnetic studies, notably those of Stanley K. Runcorn and his coworkers in England, showed that in the late Paleozoic Era the north magnetic pole—as reconstructed from European data—seems to have wandered from a Precambrian position near Hawaii to its present location in the Arctic Ocean by way of Japan. This could be explained by the migration of the magnetic pole itself (that is, polar wandering), by the migration of Europe relative to a fixed pole (that is, continental drift), or by a combination of these processes. When paleomagnetic data from other continents was obtained, each continent yielded different results. The possibility that they might reflect true wandering of the poles was discarded, because it would imply separate wanderings of many magnetic poles over the same period. However, these different paths could be reconciled into the same path by joining the continents in the manner and at the time suggested by Wegener. In other words, this analysis implied that the poles' geographic variations could be explained by the wandering of the continents. These geographic variations are called apparent polar wander paths, and they are thought to be artifacts of continental drift.

Impressed by this result, Runcorn became the first of a new generation of geologists and geophysicists to accept continental drift as a proposition worthy of careful testing. Since then, more sophisticated paleomagnetic techniques have provided both strong supporting evidence for continental drift and a major tool for reconstructing the geography and geology of the past.

Gestation and Birth of Plate-tectonic Theory

After World War II, rapid advances were made in the study of the relief, geology, and geophysics of the ocean basins. Largely because of the efforts of American oceanographer Bruce C. Heezen, American geologist Henry W. Menard, and American oceanic cartographer Marie Tharp, ocean basins, which constitute more than two-thirds of Earth's surface, became well enough known to permit serious geologic analysis. The studies revealed three very important types of features present on the ocean floor. The first type appears as broad bulges in the oceanic crust known as ocean ridges. The second set of features was revealed as deep and narrow linear troughs known as oceanic trenches. The third type occurred in seismically active fracture zones and became known as transform faults.

Seafloor spreading in three ocean basins Patterns of seafloor spreading in the Pacific (left), Arctic (centre), and Atlantic oceans (right) showing the relative age of oceanic crust. The youngest regions are coloured red, whereas the oldest regions are coloured blue.

Discovery of Ocean Basin Features

Systematic measurements of ocean depth conducted during the middle of the 20th century and the three-dimensional relief maps that were produced from these surveys revealed broad, relatively elevated oceanic ridges that form an interconnected network about 65,000 km (40,000 miles) in length and nearly girdle the globe. Ocean ridges have elevations that typically rise 2 to 3 km (1.2 to 1.9 miles) above the surrounding seafloor and widths that range from a few hundred to more than 1,000 km (600 miles). Their crests tend to be rugged and are often endowed with a rift valley at their summit where fresh lava, high heat flow, and shallow earthquakes typical of extensional environments (areas where the crust is stretched rather than compressed) are found.

These surveys also revealed long, narrow depressions—oceanic trenches—that virtually ring the Pacific Ocean; a few also occur in the northeastern part of the Indian Ocean, and some small ones are found in the central Atlantic Ocean that encircle the Caribbean Plate. Elsewhere they are absent. In contrast to ocean ridges, trenches have low heat flow, are often (but not always) filled with thick sediments, and lie at the upper edge of the Wadati-Benioff zone of compressive earthquakes. Trenches may border continents, as in the case of western Central and South America, or may occur in mid-ocean, as, for example, in the southwestern Pacific.

Offsets of up to several hundred kilometres along oceanic ridges and, more rarely, trenches were also recognized, and these fracture zones—later termed transform faults—were described as transverse features consisting of linear ridges and troughs. In oceanic domains, these faults were found to occur approximately perpendicular to the ridge crest, continue as fracture zones that extend over long distances, and terminate abruptly against continental margins. They are not sites of volcanism, and their seismic activity is restricted to the area between offset ridge crests, where earthquakes indicating horizontal slip are common.

Hess's Seafloor-spreading Model

The existence of these three types of large, striking seafloor features demanded a global rather than local tectonic explanation. The first comprehensive attempt at such an explanation was made by Harry H. Hess of the United States in a widely circulated manuscript written in 1960 but not formally published for several years. In this paper, Hess, drawing on Holmes's model of convective flow in the mantle, suggested that the oceanic ridges were the surface expressions of rising and diverging convective mantle flow, while trenches and Wadati-Benioff zones, with their associated island arcs, marked descending limbs. At the ridge crests, new oceanic crust would be generated and then carried away laterally to cool, subside, and finally be destroyed in the nearest trenches. Consequently, the age of the oceanic crust should increase with distance away from the ridge crests, and, because recycling was its ultimate fate, very old oceanic crust would not be preserved anywhere. This model explained why rocks older than 200 million years had never been encountered in the oceans, whereas the continents preserve rocks almost 4 billion years old.

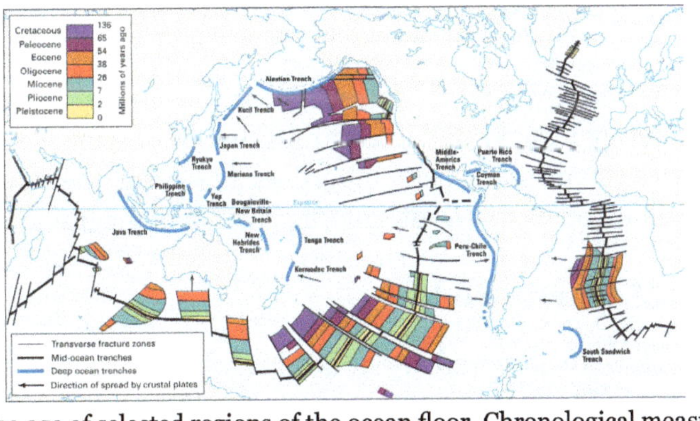

Map showing the age of selected regions of the ocean floor. Chronological measurements were derived from analysis of magnetic irregularities in the oceanic crust near spreading centres.

Hess's model was later dubbed seafloor spreading by the American oceanographer Robert S. Dietz. Confirmation of the production of oceanic crust at ridge crests and its subsequent lateral transfer came from an ingenious analysis of transform faults by Canadian geophysicist J. Tuzo Wilson. Wilson argued that the offset between two ridge crest segments is present at the outset of seafloor spreading. As each ridge segment generates new crust that moves laterally away from the ridge, the crustal slabs move in opposite directions along that part of the fracture zone that lies between the crests. In the fracture zones beyond the crests, adjacent portions of crust move in parallel (and are therefore aseismic—that is, do not have earthquakes) and are eventually consumed in a subduction zone. Wilson called this a transform fault and noted that on such a fault the seismicity should be confined to the part between ridge crests, a prediction that was subsequently confirmed by an American seismologist, Lynn R. Sykes.

The Vine-Matthews Hypothesis

Magnetic Anomalies

In 1961 a magnetic survey of the eastern Pacific Ocean floor off the coast of Oregon and California was published by two geophysicists, Arthur D. Raff and Ronald G. Mason. Unlike on the continents, where regional magnetic anomaly patterns (that is, magnetic patterns that deviate from the common rule) tend to be confused and seemingly random, the seafloor possesses a remarkably regular set of magnetic bands of alternately higher and lower values than the average values of Earth's magnetic field. These positive and negative anomalies are strikingly linear and parallel with the oceanic ridge axis, show distinct offsets along fracture zones, and generally resemble the pattern of a zebra skin. The axial anomaly tends to be higher and wider than the adjacent ones, and in most cases the sequence on one side is the approximate mirror image of that on the other.

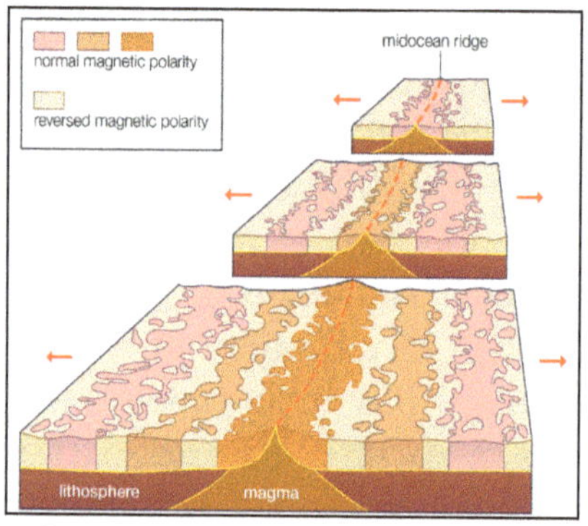

Seafloor spreading and magnetic striping.

Rising magma assumes the polarity of Earth's geomagnetic field before it solidifies into oceanic crust. At spreading centres, this crust is separated into parallel bands of rock by successive waves of emergent magma. When Earth's geomagnetic field undergoes a reversal, the change in polarity is recorded in the magma, which contributes to the alternating pattern of magnetic striping on the seafloor.

Reversal of Earth's Magnetic Field

The magnetic patterns that were observed on the seafloor defied explanation until the 1960s. Ironically, the clue for understanding these patterns came from the analysis of the magnetic properties of basaltic rocks on land. Basaltic lavas were extruded in rapid succession in a single locality on land and showed that the north and south poles had apparently repeatedly interchanged. This could be interpreted in one of two ways—either the rocks must have somehow reversed their magnetism, or the polarity of Earth's magnetic field must periodically reverse itself. Allan Cox of Stanford University and Brent Dalrymple of the United States Geological Survey collected magnetized samples around the world and showed that they displayed the same reversal at the same time, implying that the polarity of Earth's magnetic field periodically reversed. These studies established a sequence of reversals dated by isotopic methods.

Assuming that the oceanic crust is indeed made of basalt intruded in an episodically reversing geomagnetic field, Drummond H. Matthews of the University of Cambridge and a research student, Frederick J. Vine, postulated in 1963 that the new crust would have a magnetization aligned with the field at the time of its formation. If the magnetic field was normal, as it is today, the magnetization of the crust would be added to that of Earth and produce a positive anomaly. If intrusion of new magma had taken place during a period of reverse magnetic polarity, it would subtract from the present field and appear as a negative anomaly. Subsequent to intrusion, each new block created at a spreading centre would split, and the halves, in moving aside, would generate the observed bilateral magnetic symmetry. The widths of individual anomalies should correspond to the intervals between magnetic reversals.

In 1966, correlation of magnetic traverses from different oceanic ridges demonstrated an excellent correspondence with the magnetic polarity-reversal timescale established by Cox and Dalrymple on land. This reversal timescale went back some 3 million years, but since then further extrapolation based on marine magnetic anomalies (confirmed by deep-sea drilling) has extended the magnetic anomaly timescale far into the Cretaceous Period (145 million to 66 million years ago) Subsequent research in the 21st century suggests that the oldest oceanic crust may have been formed during the middle of the Jurassic Period (174.1 million to 163.5 million years ago).

About the same time, Canadian geologist Laurence W. Morley, working independently of Vine and Matthews, came to the same explanation for the marine magnetic anomalies. Publication of his paper was delayed by unsympathetic referees and technical problems and occurred long after Vine's and Matthews's work had already firmly taken root.

Toward a Unifying Theory

Determination of Plate Thickness

After marine magnetic anomalies were explained, the cumulative evidence caused the concept of seafloor spreading to be widely accepted. However, the process responsible for continental drift remained enigmatic. Two important concerns remained. The spreading seafloor was generally seen as a thin-skin process, most likely having its base at the Moho—that is, the boundary between the crust and mantle. If only oceanic crust was involved in seafloor spreading, as seemed to be the case in the Pacific Ocean, the thinness of the slab was not disturbing, even though the ever-increasing number of known fracture zones with their close spacing implied oddly narrow but very long convection cells. More troubling was the fact that the Atlantic Ocean had a well-developed oceanic ridge but lacked trenches adequate to dispose of the excess oceanic crust. This implied that the adjacent continents needed to travel with the spreading seafloor, a process that, given the thin but clearly undeformed slab, strained credulity.

Working independently but along very similar lines, Dan P. McKenzie and Robert L. Parker of Britain and W. Jason Morgan of the United States resolved these issues. McKenzie and Parker showed with a geometric analysis that, if the moving slabs of crust were thick enough to be regarded as rigid and thus to remain undeformed, their motions on a sphere would lead precisely to those divergent, convergent, and transform boundaries that are indeed observed. Morgan demonstrated that the directions and rates of movement had been faithfully recorded by magnetic anomaly patterns and transform faults. He also proposed that the plates extended approximately 100 km (60 miles) to the base of a rigid lithosphere, which coincided with the top of the weaker asthenosphere. Seismologists had previously identified this boundary, which is marked by strong attenuation of earthquake waves, as a fundamental division in Earth's upper layers. Therefore, according to Morgan, this was the boundary above which the plates moved.

In 1968, a computer analysis by the French geophysicist Xavier Le Pichon proved that the plates did indeed form an integrated system where the sum of all crust generated at oceanic ridges is balanced by the cumulative amount destroyed in all subduction zones. That same year the American geophysicists Bryan Isacks, Jack Oliver, and Lynn R. Sykes showed that the theory, which they enthusiastically labeled the "new global tectonics," could account for the larger part of Earth's seismic activity. Almost immediately, others began to consider seriously the ability of the theory to explain mountain building and sea-level changes.

Plate-driving Mechanisms and the Role of the Mantle

By the late 1960s, details of the processes of plate movement and of boundary interactions, along with much of the plate history of the Cenozoic Era (the past 66 million years), had been worked out. Yet the driving forces that bedeviled Wegener continue to remain enigmatic because there is little information about what happens beneath the plates.

Mantle Convection

Most agree that plate movement is the result of the convective circulation of Earth's heated interior, much as envisaged by Arthur Holmes in 1929. The heat source for convection is thought to be the decay of radioactive elements in the mantle. How this convection propels the plates is poorly understood. In the western Pacific Ocean the subduction of old dense oceanic crust may be self-propelled. The weight of the subducted slab may drag the rest of the plate toward the trench, a process known as slab pull, much as a tablecloth will pull itself off a table if more than half of the cloth is draped over the table's edge.

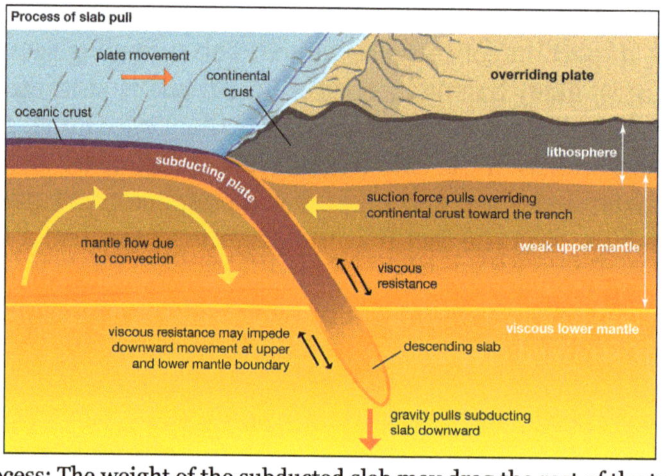

Slab pull process: The weight of the subducted slab may drag the rest of the tectonic plate toward the trench, a process known as slab pull, much as a tablecloth will pull itself off a table if more than half of the cloth is draped over the table's edge.

Some geologists argue that the westward drift of North America and eastward drift of Europe and Africa may be due to push at the spreading ridge (the Mid-Atlantic Ridge), known as ridge push, in the Atlantic Ocean. This push is caused by gravitational force, and it exists because the ridge occurs at a higher elevation than the rest of the ocean floor. As rocks near the ridge cool, they become denser, and gravity pulls them away from the ridge. As a result, new magma is allowed to well upward from the underlying hot mantle.

Hot mantle that spreads out laterally beneath the ridges or at hotspots may speed up or slow down the plates, a force known as mantle drag. However, the mantle flow pattern at depth does not appear to be reflected in the surface movements of the plates.

The relationship between the circulation within Earth's mantle and the movement of the lithospheric plates remains a first-order problem in the understanding of plate-driving mechanisms. Circulation in the mantle occurs by thermal convection, whereby warm buoyant material rises and cool dense material sinks. Convection is possible even though the mantle is solid; it occurs by solid-state creep, analogous to the slow downhill movement of valley glaciers. Materials can flow in this fashion if they are close to their melting temperatures. Several different models of mantle convection have been proposed. The simplest, called whole mantle convection, describes the presence of several large cells that rise from the core-mantle boundary beneath oceanic ridges and begin their descent to that boundary at subduction zones. Some geophysicists argue for layered mantle convection, suggesting that more vigorous convection in the upper mantle is decoupled from that in the lower mantle. This model requires that the boundary between the upper and lower mantle is coincident with a change in composition. A third model, known as the mantle plume model, suggests that upwelling is focused in plumes that ascend from the core-mantle boundary, whereas diffuse return flow is accomplished by subduction zones, which, according to this model, extend to the core-mantle boundary.

Seismic Tomography

A powerful technique, seismic tomography, provides insight into the understanding of plate-driving mechanisms. This technique is similar in principle to that of the CT (computed tomography) scan and creates three-dimensional images of Earth's interior by combining information from many earthquakes. Seismic waves generated at the site, or focus, of an earthquake spread out in all directions, similar to light rays from a light source. As earthquakes occur in many parts of Earth's crust, information from many sources can be synthesized, mimicking the rotating X-ray beam of a CT scan. Because their speed depends on the density, temperature, pressure, rigidity, and phase of the material through which they pass, the velocity of seismic waves provides clues to the composition of Earth's interior. Seismic energy is absorbed by warm material, so that the waves are slowed down. As a result, anomalously warm areas in the mantle are seismically slow, clearly distinguishing them from colder, more rigid, anomalously fast regions.

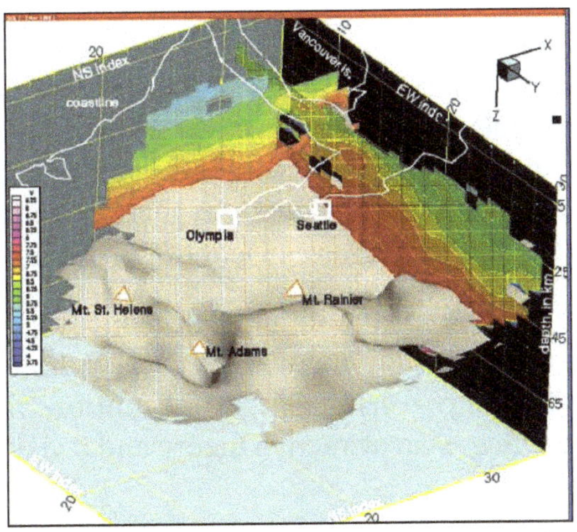

A three-dimensional slice of the geology of western Washington imaged with seismic tomography.

Tomographic imaging shows a close correspondence between surface features such as ocean ridges and subduction zones to a depth of about 100 km (60 miles). Hot regions in the mantle occur beneath oceanic ridges, and cold regions occur beneath subduction zones. However, at greater depths the pattern is more complex. Tomographic images can track the subduction zones beneath Central America and Japan close to the core-mantle boundary, suggesting that a 670-km- (about 420-mile-) deep transition between the upper and lower mantle is not an impenetrable barrier to mantle flow. These images also indicate that some subducted slabs have accumulated in slab graveyards (regions where parts of subducted slabs remain partially intact); these were observed in the mantle beneath portions of several continents and oceans but above the core-mantle boundary.

Geodynamic models suggest that subducted slabs may initially collect at a depth of 670 km beneath the surface, before rapidly descending toward the core-mantle boundary (located some 2,900 km [1,800 miles] deep) in a process known as a slab avalanche. More generally, the geochemical and isotopic compositions of oceanic basalts (which originate by melting of the mantle) appear to require a chemical contribution from the subducted slabs. Taken together, the available data indicate that subducted slabs penetrate into the deep mantle and that slab pull is an important plate-driving mechanism.

Imaging the mantle directly beneath hotspots has identified anomalously warm mantle down to the core-mantle boundary, providing strong evidence for the existence of plumes and the possibility that the mantle plume hypothesis may be a significant contributing process to mantle convection.

The composition of the deep mantle (the lowermost 300–500 km [200–300 miles] of the mantle) is considered to be very heterogeneous, and it may play a fundamental role in plate-driving mechanisms. In addition to being the potential graveyard for subducted slabs throughout much of geologic time, the heterogeneous nature of the deep mantle may be the product of chemical exchanges between the core and the mantle. Furthermore, experiments that mimic conditions near the core-mantle boundary have identified an important mineral reaction (the perovskite to post-perovskite reaction). This reaction is exothermic (it releases energy), and thus it could enhance the production of mantle plumes.

Dissenting Opinions and Unanswered Questions

Dissenters

After decades of controversy, the concept of continental drift was finally accepted by the majority of Western scientists as a consequence of plate tectonics. Sir Harold Jeffreys continued his lifelong rejection of continental drift on grounds that his estimates of the properties of the mantle indicated the impossibility of plate movements. He did not, in general, consider the mounting geophysical and geologic arguments that supported the concept of Earth's having a mobile outer shell.

Russian scientists, most notably Vladimir Vladimirovich Belousov, continued to advocate a model of Earth with stationary continents dominated by vertical motions. The model, however, only vaguely defined the forces supposedly responsible for the motions. In later years, Russian geologists came to regard plate tectonics as an attractive theory and a viable alternative to the concepts of Belousov and his followers.

In 1958 the Australian geologist S. Warren Carey proposed a rival model, known as the expanding Earth model. Carey accepted the existence and early Mesozoic breakup of Pangea and the subsequent dispersal of its fragments and formation of new ocean basins, but he attributed it all to the expansion of Earth, the planet presumably having had a much smaller diameter in the late Paleozoic Era. In his view, the continents represented the preexpansion crust, and the enlarged surface was to be entirely accommodated within the oceans. This model accounted for a spreading ocean floor and for the young age of the oceanic crust; however, it failed to deal adequately with the evidence for subduction and compression. Carey's model also did not explain why the process should not have started until some four billion years after Earth was formed, and it lacked a reasonable mechanism for so large an expansion. Finally, it disregarded the evidence for continental drift before the existence of Pangea.

Unanswered Questions

In his famous book The Structure of Scientific Revolutions (1962), the philosopher Thomas S. Kuhn pointed out that science does not always advance in the gradual and stately fashion commonly attributed to it. Most natural sciences begin with observations collected at random, without much regard to their significance or relationship between one another. As the numbers of observations increase, someone eventually synthesizes them into a comprehensive model, known as a paradigm. A paradigm is the framework, or context that is assumed to be correct, and so guides interpretations and other models. When a paradigm is accepted, advances are made by application of the paradigm. A crisis arises when the weight of observations points to the inadequacy of the old paradigm and there is no comprehensive model that can explain these contradictions. Major breakthroughs often come from an intuitive leap that may be contrary to conventional wisdom and widely accepted evidence, while strict requirements for verification and proof are temporarily relaxed. If a new paradigm is to be created, it must explain most of the observations of the old paradigm and most of the contradictions.

This paradigm shift constitutes a scientific revolution; therefore, it often becomes widely accepted before the verdict from rigorous analysis of evidence is completely in. Such was certainly the case with the geologic revolution of plate tectonics, which also confirms Kuhn's view that a new paradigm is unlikely to supersede an existing one until there is little choice but to acknowledge that the conventional theory has failed. Thus, while Wegener did not manage to persuade the scientific world of continental drift, the successor theory, plate tectonics, was readily embraced 40 years later, even though it remained open to much of the same criticism that had caused the downfall of continental drift.

The greatest successes of plate tectonics have been achieved in the ocean basins, where additional decades of effort have confirmed its postulates and enabled investigators to construct a precise and credible history of past plate movements for the past 150 million to 200 million years. Early Mesozoic and Paleozoic continental reconstructions are less rigorous because the contemporary oceanic crust has been subducted. For the most part, the principles used to reconstruct continents prior to 150 million years ago are similar to those used by Wegener (i.e., matching the geological evolution of ancient continents and terranes). However, the modern database is far greater than what was available to Wegener during his time; it has allowed for plate reconstructions with far greater resolution. Although geoscientists differ with one another over some of the details, there is a broad consensus on plate reconstructions for the Phanerozoic Eon (541 million years ago to the present).

Plate Tectonics and the Geologic Past

The extent to which plate tectonics has influenced Earth's evolution through geologic time depends on when the process started. This is a matter of ongoing debate among geologists. The principal problem is that almost all oceanic crust older than about 200 million years has been obliterated by subduction. Some of the other hallmarks of subduction—such as the high-pressure low-temperature metamorphic belts and the preservation of ophiolites—are very poorly represented in orogenic belts (belts of former or actual mountain ranges resulting from crustal deformation related to subduction or continental collision) that are older than 600 million years. To some geoscientists, this implies that tectonic processes guiding the evolution of Earth were different from those of today. Other geoscientists, however, point out that these features are unlikely to be preserved in ancient orogenic belts or that their absence may be explained by the higher geothermal gradient that must have been present during much of the Precambrian. Although thick sequences of marine sedimentary rocks up to 3.5 billion years old imply that oceanic environments did exist early in Earth's history, virtually none of the oceanic crust that underlay these sediments has been preserved. Despite these disadvantages, there is enough fragmentary evidence to suggest that plate-tectonic processes similar to those of today extend back in time at least as far as the Paleoproterozoic Era, some 2.5 billion to 1.6 billion years ago.

Wilson Cycle

The first step toward this conclusion was once again provided by Tuzo Wilson in 1966, when he proposed that the Appalachian-Caledonide mountain belt of western Europe and eastern North America was formed by the destruction of a Paleozoic ocean that predated the Atlantic Ocean. Wilson was impressed with the similarity of thick sequences of Cambrian-Ordovician marine sediments to those of modern continental shelves. In a Pangea reconstruction, Wilson showed that these shelflike sedimentary sequences extend along the entire length of the mountain chain from Scandinavia to the southeastern United States. However, these shelf sequences contain two distinct fossil assemblages on opposite sides of the mountain chain. The assemblages on the western and eastern sides of the mountain chain were named the Pacific and Atlantic realms, respectively. Both realms can be traced for thousands of kilometres along the length of the mountain belt, but neither can be traced across it. Wilson concluded that these sediments with distinctive faunal realms represented two opposing flanks of an ancient ocean that was consumed by subduction to form the Appalachian-Caledonide mountain belt. According to this model, subduction in this ocean in Ordovician times (485.4 million to 443.8 million years ago) led to the foundering of the continental shelves and the formation of volcanic arcs. The basaltic complexes of western Newfoundland were interpreted as ophiolitic complexes representing slivers of oceanic crust that escaped subduction as they were emplaced onto the continental margin. Continued subduction resulted in the closure of this ocean sometime during the Silurian and Devonian periods (443.8 million to 358.9 million years ago). Rocks representing each of these environments were found, lending strong support to this model. This ocean was subsequently named Iapetus, for the father of Atlantis in Greek mythology.

The concept that oceans may close and then reopen became known as the Wilson cycle, and with its acceptance came the application of plate-tectonic principles to ancient orogenic belts. But how far back these principles may be extended is still an open question.

In the absence of the seafloor record, evidence of ancient oceans may be obtained from thick sedimentary sequences similar to those of modern continental shelves or from preserved features formed in or around subduction zones, such as accretionary wedges, glaucophane-bearing blueschists, volcanic rocks with compositions similar to modern island arcs, and remnants of oceanic crust preserved as ophiolites obducted onto continents. These features can be confidently identified in the Paleozoic Era (approximately 541 million to about 252 million years ago) and possibly in the Neoproterozoic Era (the later part of the Proterozoic Eon, occurring approximately 1 billion to 541 million years ago), but their recognition is more problematic with increasing age. It is not known with any certainty whether this is because the dynamic nature of Earth's surface obliterates such evidence or because processes different from the modern form of plate tectonics existed at the time. Since the 1990s, however, most geoscientists have begun to accept that some form of plate tectonics occurred throughout the Proterozoic Eon, which commenced 2.5 billion years ago, and some extend these models back into the Archean, which most geologists maintain lasted from 4 billion to 2.5 billion years ago. Indeed, seismic studies across the Canadian Shield have identified buried geophysical anomalies that may represent the vestiges of subduction zones that may be 2.7 billion years old.

Some of the critical field evidence supporting the presence of subduction some 2 billion years ago comes from a suite of rocks in the Canadian Shield known as the Trans-Hudson belt. This belt separates stable regions of continental crust, known as cratons. Marc St-Onge and colleagues from the Geological Survey of Canada provided strong evidence that the formation of the Trans-Hudson belt represents the oldest documented example of a Wilson cycle in which the cratonic areas, once separated by oceans, were brought together by subduction and continental collision. They found thick sedimentary sequences typical of modern continental rifts that are about 2 billion years old, and they found ophiolites of about the same age, which indicated that rifting resulted in continental drift and formation of an ocean. About 1.85 billion years ago, volcanic rocks typical of modern island arcs were deposited on top of this sequence, indicating that the continental margins had foundered and become subduction zones. Finally, they dated the time of continental collision at about 1.8 billion years. This collisional event is particularly important because welding the cratons together provided the core of Laurentia, the continent that was ultimately to become North America.

Supercontinent Cycle

Although the Wilson cycle provided the means for recognizing the formation and destruction of ancient oceans, it did not provide a mechanism to explain why this occurred. In the early 1980s a controversial concept known as the supercontinent cycle was developed to address this problem. When viewed in a global context, it is apparent that episodes of continental rifting and mountain building are not evenly distributed throughout geologic time but instead are concentrated in relatively short time intervals approximately 350 million to 500 million years apart. Mountain building associated with the formation of Pangea peaked at about 300 million years ago. This episode was preceded by other mountain-building events peaking at 600 million to 650 million years ago and at 1.1, 1.6, 2.1, and 2.6 billion years ago. Like Pangea, could these episodes represent times of supercontinent amalgamation? Similarly, the breakup of Pangea is documented by continental-rifting events that began about 200 million years ago. However, regionally extensive and thick sequences of similar deposits occur 550 million years ago and 1, 1.5, and 2 billion years ago. Could these represent times of supercontinent dispersal?

If indeed a supercontinent cycle exists, then there must be mechanisms responsible for breakup and amalgamation. The first step is to examine why a supercontinent like Pangea would break up. There are several theories, the most popular of which, proposed by American geophysicist Don Anderson, attributes breakup to the insulating properties of the supercontinent, which blocks the escape of mantle heat. As a result, the mantle beneath the supercontinent becomes anomalously hot, and vast volumes of basaltic magma pond beneath it, forcing it to arch up and crack. Magma invades the cracks, and the process of continental rifting, ultimately leading to seafloor spreading, begins. This model implies that supercontinents have built-in obsolescence and can exist only for so long before the buildup of heat beneath them results in their fragmentation. The dating of emplacement of vast suites of basaltic magma, known as basaltic dike swarms, is consistent with the ages of continental rifting, suggesting that mantle upwelling was an important contributor to the rifting process.

The processes initiating subduction that would bring reassembled continents into a supercontinent are controversial. One theory proposes that the relative youth of modern oceanic lithosphere, which is less than 200 million years old, supports the notion that old oceanic lithosphere becomes gravitationally unstable (denser) with age and that it spontaneously subducts. Thus, as oceanic lithosphere formed by supercontinent dispersal ages, it has a tendency to subduct, possibly at fracture zones. The subducting slab undergoes mineralogical changes as it descends, resulting in slab pull that eventually hauls one section of lithosphere capped by continental crust to the subduction zone. Upon its arrival in the subduction zone, this relatively buoyant continental crust does not subduct to any appreciable degree. Instead, it collides with other masses of continental crust located behind the subduction zone and contributes to the formation of a new supercontinent.

Continental Reconstructions

Magnetic anomalies, transform faults, hotspots, and apparent polar wandering paths permit rigorous geometric reconstructions of past plate positions, shapes, and movements. Although some important controversies remain, these paleogeographic reconstructions show the changing geography of Earth's past and can be determined with excellent precision for the past 150 million years. Before that time, however, the absence of the ocean-floor record makes the process significantly more challenging. A variety of geologic data are used to help determine the proper fit of continents through time. Some of the methods used to test these reconstructions are based on matching patterns from one continental block to another and are similar to the approach of Wegener. However, modern geoscientists have more precise data that help constrain these reconstructions. Of the many advances, perhaps the most significant are the improved analytical techniques for radiometric dating, allowing the age of geologic events to be determined with much greater precision. One of the most common methods used measures the radioactive decay of uranium to lead in the mineral zircon by comparing the ratio of one to the other in the sample of zircon. Zircon is a common accessory mineral in igneous, metamorphic, and sedimentary rocks. Modern techniques typically yield age determinations with an estimated error of 2 million years or less, even for rocks of Archean age.

Since the 1990s the database has improved so that reasonably constrained reconstructions can now be made as far back as 1 billion years. For example, the abundance of continental-collisional events about 1.1 billion years ago is one of the principal lines of evidence suggesting the presence of a supercontinent that is given the name of Rodinia. By about 760 million years ago a number of continental-rift sequences had developed, suggesting that Rodinia had begun to break up. Between about 650 million and 550 million years ago, however, a number of mountain belts formed by continental collision, which resulted in the amalgamation of Gondwana, the supercontinent originally identified by Du Toit in 1937. The continental fragment that rifted away from Laurentia did not return to collide with North America as predicted by a simple Wilson cycle. Instead, it rotated counterclockwise away from Laurentia until it collided with eastern Africa.

Interactions of Tectonics with Other Systems

Sea Level

As plate tectonics changes the shape of ocean basins, it fundamentally affects long-term variations in global sea level. For example, the geologic record in which thick sequences of continental shelf sediments were deposited demonstrates that the breakup of Pangea resulted in the flooding of continental margins, indicating a rise in sea level. There are several contributing factors. First, the presence of new ocean ridges displaces seawater upward and outward across the continental margins. Second, the dispersing continental fragments subside as they cool. Third, the volcanism associated with breakup introduces greenhouse gases in the atmosphere, which results in global warming, causing continental glaciers to melt.

As an ocean widens, its crust becomes older and denser. It therefore subsides, eventually forming ocean trenches. As a result, ocean basins can hold more water, and sea level drops. This changes once again when subduction commences. Subduction preferentially consumes the oldest oceanic crust, so that the average age of oceanic crust becomes younger. Younger oceanic crust is therefore more buoyant and has a higher elevation, a circumstance that causes sea level to rise once more.

Composition of Ocean Water

Water's strong properties as a solvent mean that it is rarely pure. Ocean water contains about 96.5 percent by weight pure water, with the remaining 3.5 percent predominantly consisting of ions such as chloride (1.9 percent), sodium, (1.1 percent), sulfate (0.3 percent), and magnesium (0.1 percent). The drainage of water from continents via the hydrologic cycle plays an important role in transporting chemicals from the land to the sea. The effect of this drainage is profoundly influenced by the presence of mountain belts. For example, the erosional power of the Ganges River, which drains from the Himalayas, carries 1.45 billion metric tons of sediment to the sea annually. This load is nine times that of the Mississippi River. The processes of weathering and erosion strip soluble elements such as sodium from their host minerals, and the relatively high concentration of sodium in ocean water is attributed to the weathering and erosion that accompanies continental drainage.

Until the advent of plate tectonics, discovering the source of chlorine was problematic because chlorine is present in only very minor amounts in the continental crust. Scientists hypothesized that the source of this element may lie in underwater volcanic activity. In the late 1970s three scientists investigating the oceanic ridge off the coast of Peru from a submersible craft documented the occurrence of superheated jets of water, up to 350 °C (660 °F), continuously erupting from chimneys that stood 13 metres (43 feet) above the ocean floor. These hot springs were found to be rich in chlorine and metals, confirming that the source of chlorine in the oceans lay in the tectonic processes occurring at oceanic ridges.

Life

Plate tectonics has influenced the evolution and propagation of life in a variety of ways. The study of oceanic ridges revealed the presence of bizarre life adjacent to the chimneys of superheated water that together makes up about 1 percent of the world's ecosystems. The existence of these life-forms in the deep ocean cannot be based on photosynthesis. Instead, they are nourished by minerals and heat. The energy released when hydrogen sulfide in the vent reacts with seawater is used by bacteria to convert inorganic carbon dioxide dissolved in seawater into organic compounds, a process known as chemosynthesis. Some scientists speculate that the cumulative influence of this process over time has had a significant effect on evolution. Others suggest that similar processes may ultimately be responsible for the origin of life on Earth.

Vent tube worms (Riftia pachyptila).

The continuous rearrangement of the size and shape of ocean basins and continents over geologic time, accompanied by changes in ocean circulation and climate, had a major impact on the development of life on Earth. One of the first studies of the potential effects of plate tectonics on life was published in 1970 by American geologists James W. Valentine and Eldridge M. Moores, who proposed that the diversity of life increased as continents fragmented and dispersed and diversity diminished when the continents were joined together.

Evolution

When Pangea began to rift apart and the Atlantic Ocean started to open during the middle Mesozoic, the differences between the faunas of opposite shores gradually increased in an almost linear fashion—the greater the distance, the smaller the number of families in common. The difference increased more rapidly in the South Atlantic than in the North Atlantic, where a land connection between Europe and North America persisted until about 60 million years ago.

After the breakup of Pangea, no land animal or group of animals could become dominant, because the continents were disconnected. As a result, the separated landmasses evolved highly specialized fauna. South America, for example, was rich in marsupial mammals, which had few predators. North America, on the other hand, was rich in placental mammals.

opossum
Nocturnal animals, such as opossums, have eyes with large, nearly spherical lenses.

However, about 3 million years ago, volcanic activity associated with subduction of the eastern Pacific Ocean formed a land bridge across the isthmus of Panama, reconnecting the separate landmasses. The emergence of the isthmus made it possible for land animals to cross, forcing previously separated fauna to compete. Numerous placental mammals and herbivores migrated from north to south. They adapted well to the new environment and were more successful than the local fauna in competing for food. The invasion of highly adaptable carnivores from the north contributed to the extinction of at least four orders of South American land mammals. A few species, notably the armadillo and the opossum, managed to migrate in the opposite direction. Ironically, many of the invading northerners, such as the llama and tapir, subsequently became extinct in their region of origin and found their last refuge to the south.

Extinction

Perhaps the most dramatic example of the potential impact of plate tectonics on life occurred near the end of the Permian Period (roughly 299 million to 252 million years ago). Several events contributed to the Permian extinction that caused the permanent disappearance of half of Earth's known biological families. The marine realm was most affected, losing more than 90 percent of its species. About 70 percent of terrestrial species became extinct. This extinction appears to have occurred in several pulses, and there may have been numerous contributing factors—including biogeographic changes associated with the formation of Pangea (which would have been accompanied by a sharp decrease in area of shallow-water habitats), changes in the patterns of nutrient-rich deep ocean currents, changes in the amount of dissolved oxygen in ocean waters, and temperature increases and changes to the carbon cycle caused partly by the population explosion of the methane-producing microbe Methanosarcina. Another contributing factor could have been the environmental consequences of the vast volcanic outpourings of the Siberian Traps, one of the largest volcanic events documented. It produced a region of flood basalt that had an estimated volume of 2,000,000–3,000,000 million cubic km [about 480,000–720,000 cubic miles]). The Siberian Traps eruption occurred about the same time as the extinction, and the greenhouse gases emitted from these volcanoes may have affected the amount of acidity of the oceans.

The extinction had a complex history. High latitudes were affected first as a result of the waning of the Permian ice age when the southern edge of Pangea moved off the South Pole. The equatorial and subtropical zones appear to have been affected somewhat later by a global cooling. On the other hand, the extinctions were not felt as strongly on the continent itself. Instead, the vast semiarid and arid lands that emerged on so large a continent, the shortening of its moist coasts, and the many mountain ranges formed from the collisions that led to the formation of the supercontinent provided strong incentives for evolutionary adaptation to dry or high-altitude environments.

Climate

Climate changes associated with the supercontinent of Pangea and with its eventual breakup and dispersal provide an example of the effect of plate tectonics on paleoclimate. Pangea was completely surrounded by a world ocean (Panthalassa) extending from pole to pole and spanning 80 percent of the circumference of Earth at the paleoequator. The equatorial current system, driven by the trade winds, resided in warm latitudes much longer than today, and its waters were therefore warmer. The gyres that occupy most of the Southern and Northern hemispheres were also warmer, and consequently the temperature gradient from the paleoequator to the poles was less pronounced than it is at present.

Early in the Mesozoic Era, Gondwana split from its northern counterpart, Laurasia, to form the Tethys seaway, and the equatorial current became circumglobal. Equatorial surface waters were then able to circumnavigate the world and became even warmer. How this flow influenced circulation at higher latitudes is unclear. From about 100 million to 70 million years ago, isotopic records show, Arctic and Antarctic surface water temperatures were at or above 10 °C (50 °F), and the polar regions were warm enough to support forests.

As the dispersal of continents following the breakup of Pangea continued, however, the surface circulation of the oceans began to approach the more complex circulation patterns of today. About 100 million years ago, the northward drift of Australia and South America created a new circumglobal seaway around Antarctica, which remained centred on the South Pole. A vigorous circum-Antarctic current developed, isolating the southern continent from the warmer waters to the north. At the same time, the equatorial current system became blocked, first in the Indo-Pacific region and next in the Middle East and eastern Mediterranean and, about 6 million years ago, by the emergence of the Isthmus of Panama. As a result, the equatorial waters were heated less, and the midlatitude ocean gyres were not as effective in keeping the high-latitude waters warm. Because of this, an ice cap began to form on Antarctica some 20 million years ago and grew to roughly its present size about 5 million years later. This ice cap cooled the waters of the adjacent ocean to such a low temperature that the waters sank and initiated the north-directed abyssal flow that marks the present deep circulation.

Also at about 6 million years ago, the collision between Africa and Europe temporarily closed the Strait of Gibraltar, isolating the Mediterranean Sea and restricting its circulation. Evaporation, which produced thick salt deposits, virtually dried up this sea and lowered the salt content of the world's oceans, allowing seawater to freeze at higher temperatures. As a result, polar ice sheets grew and sea level fell. About 500,000 years later, the barrier between the Mediterranean and the Atlantic Ocean was breached, and open circulation resumed.

The Quaternary Ice Age arrived in full when the first ice caps appeared in the Northern Hemisphere about 2 million years ago. It is highly unlikely that the changing configuration of continents and oceans can be held solely responsible for the onset of the Quaternary Ice Age, even if such factors as the drift of continents across the latitudes (with the associated changes in vegetation) and reflectivity (albedo) for solar heat are included. There can be little doubt, however, that it was a contributing factor and that recognition of its role has profoundly altered concepts of paleoclimatology.

Volcanology

Volcanology or vulcanology is a discipline of the geologic sciences that is concerned with all aspects of volcanic phenomena.

Volcanology deals with the formation, distribution, and classification of volcanoes as well as with their structure and the kinds of materials ejected during an eruption (such as pyroclastic flows, lava, dust, ash, and volcanic gases). It also involves research on the relationships between volcanic eruptions and other large-scale geologic processes such as plate tectonics, mountain building, and earthquakes. One of the chief objectives of this research is determining the nature and causes of volcanic eruptions for the purpose of forecasting their occurrence. Another practical concern of volcanology is securing data that may aid in locating commercially valuable deposits of ores, particularly those of certain sulfide minerals.

Interest in volcanic phenomena extends back to ancient times. The eruption of Mount Vesuvius in AD 79 was recorded in considerable detail by Pliny the Younger. Studies of volcanoes, however, were not conducted systematically until the 19th century. Since that time volcanology has become an important branch of physical geology. Specialists in the field, using the principles and methods of geophysics and geochemistry and the tools of seismology and geodesy, have obtained much knowledge of processes that occur within the Earth's interior.

Soil Science

Soil science is the study of soil as a natural resource on the surface of the Earth including soil formation, classification and mapping; physical, chemical, biological, and fertility properties of soils; and these properties in relation to the use and management of soils.

Sometimes terms which refer to branches of soil science, such as pedology (formation, chemistry, morphology, and classification of soil) and edaphology (how soils interact with living things, especially plants), are used as if synonymous with soil science. The diversity of names associated with this discipline is related to the various associations concerned. Indeed, engineers, agronomists, chemists, geologists, physical geographers, ecologists, biologists, microbiologists, silviculturists, sanitarians, archaeologists, and specialists in regional planning, all contribute to further knowledge of soils and the advancement of the soil sciences.

Soil scientists have raised concerns about how to preserve soil and arable land in a world with a growing population, possible future water crisis, increasing per capita food consumption, and land degradation.

Fields of Study

Soil occupies the pedosphere, one of Earth's spheres that the geosciences use to organize the Earth conceptually. This is the conceptual perspective of pedology and edaphology, the two main branches of soil science. Pedology is the study of soil in its natural setting. Edaphology is the study of soil in relation to soil-dependent uses. Both branches apply a combination of soil physics, soil chemistry, and soil biology. Due to the numerous interactions between the biosphere, atmosphere and hydrosphere that are hosted within the pedosphere, more integrated, less soil-centric concepts are also valuable. Many concepts essential to understanding soil come from individuals not identifiable strictly as soil scientists. This highlights the interdisciplinary nature of soil concepts.

Research

Dependence on and curiosity about soil, exploring the diversity and dynamics of this resource continues to yield fresh discoveries and insights. New avenues of soil research are compelled by a need to understand soil in the context of climate change, greenhouse gases, and carbon sequestration. Interest in maintaining the planet's biodiversity and in exploring past cultures has also stimulated renewed interest in achieving a more refined understanding of soil.

Mapping

Most empirical knowledge of soil in nature comes from soil survey efforts. Soil survey, or soil mapping, is the process of determining the soil types or other properties of the soil cover over a landscape, and mapping them for others to understand and use. It relies heavily on distinguishing the individual influences of the five classic soil forming factors. This effort draws upon geomorphology, physical geography, and analysis of vegetation and land-use patterns. Primary data for the soil survey are acquired by field sampling and supported by remote sensing.

Classification

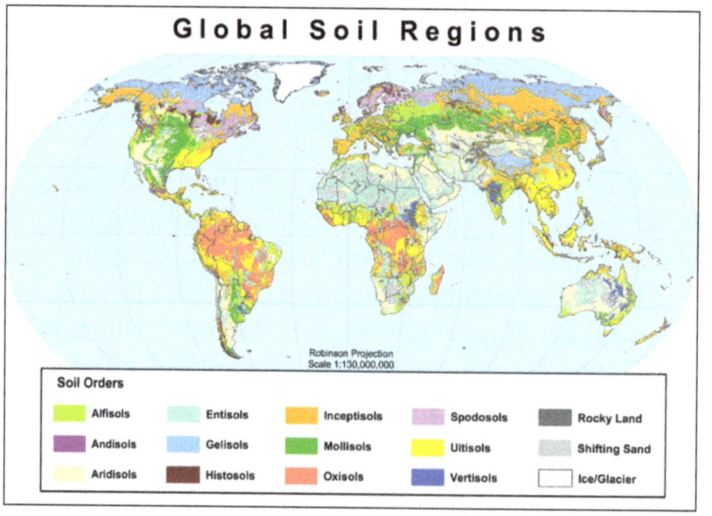

Map of global soil regions from the USDA.

As of 2006, the World Reference Base for Soil Resources, via its Land & Water Development division, is the pre-eminent soil classification system. It replaces the previous FAO soil classification.

The WRB borrows from modern soil classification concepts, including USDA soil taxonomy. The classification is based mainly on soil morphology as an expression pedogenesis. A major difference with USDA soil taxonomy is that soil climate is not part of the system, except insofar as climate influences soil profile characteristics.

Many other classification schemes exist, including vernacular systems. The structure in vernacular systems are either nominal, giving unique names to soils or landscapes, or descriptive, naming soils by their characteristics such as red, hot, fat, or sandy. Soils are distinguished by obvious characteristics, such as physical appearance (e.g., color, texture, landscape position), performance (e.g., production capability, flooding), and accompanying vegetation. A vernacular distinction familiar to many is classifying texture as heavy or light. Light soil content and better structure, take less effort to turn and cultivate. Contrary to popular belief, light soils do not weigh less than heavy soils on an air dry basis nor do they have more porosity.

Depression Storage Capacity

Depression storage capacity, in soil science, is the ability of a particular area of land to retain water in its pits and depressions, thus preventing it from flowing. Depression storage capacity, along with infiltration capacity, is one of the main factors involved in Horton overland flow, whereby water volume surpasses both infiltration and depression storage capacity and begins to flow horizontally across land, possibly leading to flooding and soil erosion. The study of land's depression storage capacity is important in the fields of geology, ecology, and especially hydrology.

Applied Geology

Applied Geology is a multidisciplinary subject that interacts with other disciplines, such as mineralogy, petrology, structural geology, hydrogeology, seismic engineering, rock engineering, soil mechanics, geophysics, remote sensing (RS-GIS-GPS), environmental geology, etc.

Petroleum Geology

Petroleum geology is the study of origin, occurrence, movement, accumulation, and exploration of hydrocarbon fuels. It refers to the specific set of geological disciplines that are applied to the search for hydrocarbons (oil exploration).

Sedimentary Basin Analysis

Petroleum geology is principally concerned with the evaluation of seven key elements in sedimentary basins:

- Source

- Reservoir
- Seal
- Trap
- Timing
- Maturation
- Migration

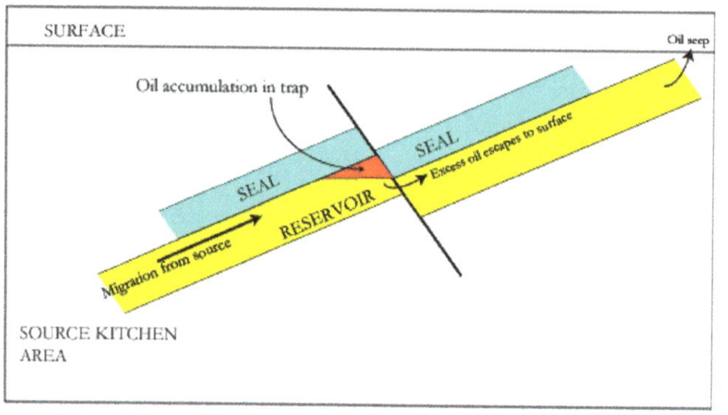

A structural trap, where a fault has juxtaposed a porous and permeable reservoir against an impermeable seal. Oil (shown in red) accumulates against the seal, to the depth of the base of the seal. Any further oil migrating in from the source will escape to the surface and seep.

In general, all these elements must be assessed via a limited 'window' into the subsurface world, provided by one (or possibly more) exploration wells. These wells present only a 1-dimensional segment through the Earth and the skill of inferring 3-dimensional characteristics from them is one of the most fundamental in petroleum geology. Recently, the availability of inexpensive, high quality 3D seismic data (from reflection seismology) and data from various electromagnetic geophysical techniques (such as Magnetotellurics) has greatly aided the accuracy of such interpretation. The following section discusses these elements in brief. For a more in-depth treatise.

Evaluation of the source uses the methods of geochemistry to quantify the nature of organic-rich rocks which contain the precursors to hydrocarbons, such that the type and quality of expelled hydrocarbon can be assessed.

The reservoir is a porous and permeable lithological unit or set of units that holds the hydrocarbon reserves. Analysis of reservoirs at the simplest level requires an assessment of their porosity (to calculate the volume of in situ hydrocarbons) and their permeability (to calculate how easily hydrocarbons will flow out of them). Some of the key disciplines used in reservoir analysis are the fields of structural analysis, stratigraphy, sedimentology, and reservoir engineering.

The seal, or cap rock, is a unit with low permeability that impedes the escape of hydrocarbons from the reservoir rock. Common seals include evaporites, chalks and shales. Analysis of seals involves assessment of their thickness and extent, such that their effectiveness can be quantified.

The trap is the stratigraphic or structural feature that ensures the juxtaposition of reservoir and seal such that hydrocarbons remain trapped in the subsurface, rather than escaping (due to their natural buoyancy) and being lost.

Analysis of maturation involves assessing the thermal history of the source rock in order to make predictions of the amount and timing of hydrocarbon generation and expulsion.

Finally, careful studies of migration reveal information on how hydrocarbons move from source to reservoir and help quantify the source (or kitchen) of hydrocarbons in a particular area.

Mud log in process, a common way to study the lithology when drilling oil wells.

Major Subdisciplines in Petroleum Geology

Several major subdisciplines exist in petroleum geology specifically to study the seven key elements discussed above.

Source Rock Analysis

In terms of source rock analysis, several facts need to be established. Firstly, the question of whether there actually is any source rock in the area must be answered. Delineation and identification of potential source rocks depends on studies of the local stratigraphy, palaeogeography and sedimentology to determine the likelihood of organic-rich sediments having been deposited in the past.

If the likelihood of there being a source rock is thought to be high, the next matter to address is the state of thermal maturity of the source, and the timing of maturation. Maturation of source rocks depends strongly on temperature, such that the majority of oil generation occurs in the 60° to 120 °C range. Gas generation starts at similar temperatures, but may continue up beyond this range, perhaps as high as 200 °C. In order to determine the likelihood of oil/gas generation, therefore, the thermal history of the source rock must be calculated. This is performed with a combination of geochemical analysis of the source rock (to determine the type of kerogens present and their maturation characteristics) and basin modelling methods, such as back-stripping, to model the thermal gradient in the sedimentary column.

Basin Analysis

A full scale basin analysis is usually carried out prior to defining leads and prospects for future drilling. This study tackles the petroleum system and studies source rock (presence and quality); burial history; maturation (timing and volumes); migration and focus; and potential regional seals and major reservoir units (that define carrier beds). All these elements are used to investigate where potential hydrocarbons might migrate towards. Traps and potential leads and prospects are then defined in the area that is likely to have received hydrocarbons.

Exploration Stage

Although a basin analysis is usually part of the first study a company conducts prior to moving into an area for future exploration, it is also sometimes conducted during the exploration phase. Exploration geology comprises all the activities and studies necessary for finding new hydrocarbon occurrence. Usually seismic (or 3D seismic) studies are shot, and old exploration data (seismic lines, well logs, reports) are used to expand upon the new studies. Sometimes gravity and magnetic studies are conducted, and oil seeps and spills are mapped to find potential areas for hydrocarbon occurrences. As soon as a significant hydrocarbon occurrence is found by an exploration- or wildcat-well the appraisal stage starts.

Appraisal Stage

The Appraisal stage is used to delineate the extent of the discovery. Hydrocarbon reservoir properties, connectivity, hydrocarbon type and gas-oil and oil-water contacts are determined to calculate potential recoverable volumes. This is usually done by drilling more appraisal wells around the initial exploration well. Production tests may also give insight in reservoir pressures and connectivity. Geochemical and petrophysical analysis gives information on the type (viscosity, chemistry, API, carbon content, etc.) of the hydrocarbon and the nature of the reservoir (porosity, permeability, etc.).

Production Stage

After a hydrocarbon occurrence has been discovered and appraisal has indicated it is a commercial find the production stage is initiated. This stage focuses on extracting the hydrocarbons in a controlled way (without damaging the formation, within commercial favorable volumes, etc.). Production wells are drilled and completed in strategic positions. 3D seismic is usually available by this stage to target wells precisely for optimal recovery. Sometimes enhanced recovery (steam injection, pumps, etc.) is used to extract more hydrocarbons or to redevelop abandoned fields.

Reservoir Analysis

The existence of a reservoir rock (typically, sandstones and fractured limestones) is determined through a combination of regional studies (i.e. analysis of other wells in the area), stratigraphy and sedimentology (to quantify the pattern and extent of sedimentation) and seismic interpretation. Once a possible hydrocarbon reservoir is identified, the key physical characteristics of a reservoir that are of interest to a hydrocarbon explorationist are its bulk rock volume, net-to-gross ratio, porosity and permeability.

Bulk rock volume, or the gross rock volume of rock above any hydrocarbon-water contact, is determined by mapping and correlating sedimentary packages. The net-to-gross ratio, typically estimated from analogues and wireline logs, is used to calculate the proportion of the sedimentary packages that contains reservoir rocks. The bulk rock volume multiplied by the net-to-gross ratio gives the net rock volume of the reservoir. The net rock volume multiplied by porosity gives the total hydrocarbon pore volume i.e. the volume within the sedimentary package that fluids (importantly, hydrocarbons and water) can occupy. The summation of these volumes for a given exploration prospect will allow explorers and commercial analysts to determine whether a prospect is financially viable.

Traditionally, porosity and permeability were determined through the study of drilling samples, analysis of cores obtained from the wellbore, examination of contiguous parts of the reservoir that outcrop at the surface and by the technique of formation evaluation using wireline tools passed down the well itself. Modern advances in seismic data acquisition and processing have meant that seismic attributes of subsurface rocks are readily available and can be used to infer physical/sedimentary properties of the rocks themselves.

Engineering Geology

Engineering geology is the application of the geology to engineering study for the purpose of assuring that the geological factors regarding the location, design, construction, operation and maintenance of engineering works are recognized and accounted for. Engineering geologists provide geological and geotechnical recommendations, analysis, and design associated with human development and various types of structures. The realm of the engineering geologist is essentially in the area of earth-structure interactions, or investigation of how the earth or earth processes impact human made structures and human activities.

Engineering geology studies may be performed during the planning, environmental impact analysis, civil or structural engineering design, value engineering and construction phases of public and private works projects, and during post-construction and forensic phases of projects. Works completed by engineering geologists include; geological hazard assessments, geotechnical, material properties, landslide and slope stability, erosion, flooding, dewatering, and seismic investigations, etc. Engineering geology studies are performed by a geologist or engineering geologist that is educated, trained and has obtained experience related to the recognition and interpretation of natural processes, the understanding of how these processes impact human made structures (and vice versa), and knowledge of methods by which to mitigate against hazards resulting from adverse natural or human made conditions. The principal objective of the engineering geologist is the protection of life and property against damage caused by various geological conditions.

The practice of engineering geology is also very closely related to the practice of geological engineering and geotechnical engineering. If there is a difference in the content of the disciplines, it mainly lies in the training or experience of the practitioner.

The Practice

One of the most important roles as an engineering geologist is the interpretation of landforms and earth processes to identify potential geologic and related man-made hazards that may have a great impact on civil structures and human development. The background in geology provides the engineering geologist with an understanding of how the earth works, which is crucial minimizing earth related hazards. Most engineering geologists also have graduate degrees where they have gained specialized education and training in soil mechanics, rock mechanics, geotechnics, groundwater, hydrology, and civil design. These two aspects of the engineering geologists' education provide them with a unique ability to understand and mitigate for hazards associated with earth-structure interactions.

Scope of Studies

Engineering geology investigation and studies may be performed:

- For residential, commercial and industrial developments;
- For governmental and military installations;
- For public works such as a stormwater drainage system, power plant, wind turbine, transmission line, sewage treatment plant, water treatment plant, pipeline (aqueduct, sewer, outfall), tunnel, trenchless construction, canal, dam, reservoir, building foundation, railroad, transit, highway, bridge, seismic retrofit, power generation facility, airport and park;
- For mine and quarry developments, mine tailing dam, mine reclamation and mine tunneling;
- For wetland and habitat restoration programs;
- For government, commercial, or industrial hazardous waste remediation sites;
- For coastal engineering, sand replenishment, bluff or sea cliff stability, harbor, pier and waterfront development;
- For offshore outfall, drilling platform and sub-sea pipeline, sub-sea cable; and
- For other types of facilities.

Geohazards and Adverse Geological Conditions

Typical geologic hazards or other adverse conditions evaluated and mitigated by an engineering geologist include:

- Fault rupture on seismically active faults ;
- Seismic and earthquake hazards (ground shaking, liquefaction, lurching, lateral spreading, tsunami and seiche events);
- Landslide, mudflow, rockfall, debris flow, and avalanche hazards ;
- Unstable slopes and slope stability;
- Erosion;
- Slaking and heave of geologic formations, such as frost heaving;
- Ground subsidence (such as due to ground water withdrawal, sinkhole collapse, cave collapse, decomposition of organic soils, and tectonic movement);
- Volcanic hazards (volcanic eruptions, hot springs, pyroclastic flows, debris flow, debris avalanche, gas emissions, volcanic earthquakes);
- Non-rippable or marginally rippable rock requiring heavy ripping or blasting;

- Weak and collapsible soils, foundation bearing failures;
- Shallow ground water/seepage;
- Other types of geologic constraints.

An engineering geologist or geophysicist may be called upon to evaluate the excavatability (i.e. rippability) of earth (rock) materials to assess the need for pre-blasting during earthwork construction, as well as associated impacts due to vibration during blasting on projects.

Soil and Rock Mechanics

Soil mechanics is a discipline that applies principles of engineering mechanics, e.g. kinematics, dynamics, fluid mechanics, and mechanics of material, to predict the mechanical behaviour of soils. Rock mechanics is the theoretical and applied science of the mechanical behaviour of rock and rock masses; it is that branch of mechanics concerned with the response of rock and rock masses to the force-fields of their physical environment. The fundamental processes are all related to the behaviour of porous media. Together, soil and rock mechanics are the basis for solving many engineering geology problems.

Methods and Reporting

The methods used by engineering geologists in their studies include:

- Geologic field mapping of geologic structures, geologic formations, soil units and hazards;
- The review of geologic literature, geologic maps, geotechnical reports, engineering plans, environmental reports, stereoscopic aerial photographs, remote sensing data, Global Positioning System (GPS) data, topographic maps and satellite imagery;
- The excavation, sampling and logging of earth/rock materials in drilled borings, backhoe test pits and trenches, fault trenching, and bulldozer pits;
- Geophysical surveys (such as seismic refraction traverses, resistivity surveys, ground penetrating radar (GPR) surveys, magnetometer surveys, electromagnetic surveys, high-resolution sub-bottom profiling, and other geophysical methods);
- Deformation monitoring as the systematic measurement and tracking of the alteration in the shape or dimensions of an object as a result of the application of stress to it manually or with an automatic deformation monitoring system;
- Other methods.

The fieldwork is typically culminated in analysis of the data and the preparation of an engineering geologic report, geotechnical report or design brief, fault hazard or seismic hazard report, geophysical report, ground water resource report or hydrogeologic report. The engineering geology report can also be prepared in conjunction with a geotechnical report, but commonly provides the same geotechnical analysis and design recommendations that would be presented in a geotechnical report. An engineering geology report describes the objectives, methodology, references cited, tests performed, findings and recommendations for development and detailed design of engineering works. Engineering geologists also provide geologic data on topographic maps, aerial photographs, geologic maps, Geographic Information System (GIS) maps, or other map bases.

References

- Outcome-what-is-geology, chapter, wmopen-geology: lumenlearning.com, Retrieved 30 April, 2019
- B.a. van der pluijm and s. Marshak (2004). Earth structure - an introduction to structural geology and tectonics (2nd ed.). New york: w. W. Norton. P. 656. Isbn 0-393-92467-x
- Plate-tectonics, science: britannica.com, Retrieved 30 August, 2019
- Leveque, randall j., 2002, finite volume methods for hyperbolic problems,cambridge university press, aug 26, 2002 isbn 0521009243
- Outcome-what-is-geology, chapter, wmopen-geology: lumenlearning.com, Retrieved 15 July, 2019
- "paleontology | science". Encyclopædia britannica. Archived from the original on 2017-08-24. Retrieved 2017-08-24
- Volcanology, science: .britannica.com, Retrieved 16 January, 2019
- H. H. Janzen; et al. (2011). "global prospects rooted in soil science". Soil science society of america journal. 75 (1): 1. Bibcode:2011ssasj..75....1j. Doi:10.2136/sssaj2009.0216
- Pielke, roger (12 december 2005). "is soil an important component of the climate system?". The climate science weblog. Archived from the originalon 8 september 2006. Retrieved 19 april 2012
- Prof. D. Venkat reddy, nit-karnataka, engineering geology, vikas publishers, 2010 isbn 978-81259-19032

Atmospheric Science 3

Atmospheric sciences deal with the study of the Earth's atmosphere. They also deal with the study of its processes, the effects other systems have on the atmosphere, as well as the effects which the atmosphere has on these systems. Some of the major atmospheric sciences are climatology, meteorology and hydrometeorology. This chapter discusses these atmospheric sciences in detail.

The study of the Earth's atmosphere and various phenomena related to it is called atmospheric science. Atmospheric science is the interdisciplinary study which combines the different components of chemistry and physics that focuses on the dynamics and structure of the atmosphere of Earth. Atmospheric science includes the study of composition, circulation, and the chemical and physical processes of the atmosphere. Atmospheric science focuses on atmosphere, the atmospheric processes, the effects of numerous systems on the atmosphere and the effects the atmosphere has on these systems. Atmospheric science extends to planetary science and studying the atmospheres of the different planets in the solar system.

Paleoclimatology

Paleoclimatology is the study of ancient climatic changes. Since going back in time to observe the climatic changes is impossible, scientists use numerous climate imprints created in the past, referred to as proxies, to interpret the paleoclimate. Some of the most reliable proxies include microfossils, shells, rocks, corals, ice sheets and tree rings among others. Scientists reconstruct the ancient climate using a combination of different categories of proxy records. The proxy records are incorporated with the observations of the current climate and then uploaded into a computer model which infers the ancient climate while predicting future climatic changes. Studies of ancient environmental changes and biodiversity always reflect on the present situation especially the impact of the climatic changes on biotic recovery and mass extinctions. Paleoclimatology started in the early 19th century when numerous discoveries about the glaciations and natural changes in the ancient climate helped scientists comprehend the greenhouse effect. The first observations with reliable scientific basis were the one observed in New Zealand by John Hardcastle in 1880s. Hardcastle discovered that the loess deposited at Timaru helped record climatic changes. Hardcastle referred to the loess as "climate registers".

Atmospheric Chemistry

Atmospheric chemistry is the field of atmospheric science which studies the chemistry of the atmosphere of earth and the other planets. Atmospheric chemistry is a multi-disciplinary approach to research which draws from volcanology, geology, environmental chemistry, meteorology, oceanography and computer modeling. The atmospheric chemistry and composition are crucial for numerous reasons, one of them being the interactions between the all the living organisms and the atmosphere. Multiple natural processes including lighting and volcano emission change the composition of the atmosphere. Atmospheric chemistry has addressed numerous problems including acid rain, global warming, photochemical smog, ozone depletion, and greenhouse gases. The atmospheric chemist tries to understand the causes of these issues and obtain a theoretical understanding of the problem which helps them create a solution which is tested and implemented.

Atmospheric Physics

Atmospheric physics is the use of physics when studying the atmosphere. Atmospheric physicists try to model the atmosphere of the earth among other planets using numerous fluid flow equations, radiation budgets, energy transfers, and chemical models. Atmospheric physics is closely related to climatology and meteorology, plus it covers the construction and design of the instruments used in studying the atmosphere and interpretation of the collected data. The atmospheric physicists use the elements of scattering theory, cloud physics, spatial statistics, and wave propagation models including the remote sensing instruments to model the weather systems. The introduction of the sounding rockets saw aeronomy become a sub-discipline dealing with the top layer of the atmosphere.

Paleotempestology

Emanuel Kerry coined the term Paleotempestology. Paleotempestology refers to the study of ancient tropical cyclone activities using numerous geological proxies and documented historical records. Some of the most efficient paleotempestology methods include sedimentary proxy records, makers in coral, historical records, and tree rings and speleothems.The sedimentary proxy records method uses the over-wash deposits conserved on the sediments of marshes, microfossils and coastal lakes. The scientists adopted the use of over-wash deposits from the earlier studies of numerous paleotsunami deposits. The first study of a cyclone occurred in South Pacific and Australia from the late 1970s to early 1980s. The studies examined many parallel coral shingle ridges and marine shells and confirmed that cyclones deposit over 50 ridges on the site, and each represents an ancient severe cyclone which occurred thousands of years ago. Rocks have some natural isotopes of elements referred to as natural tracers, which help describe the state under which the rock formed. Studying the calcium carbonate present in the coral rocks helps reveal the hurricane information and surface temperature of when it developed. The heavy oxygen isotopes decrease faster as compared to lighter oxygen isotopes during the heavy rainfall periods. Since the hurricanes were the primary sources of heavy rainfall in tropical oceans, scientists can date the ancient storms by looking at the decreased lighter oxygen isotope in the coral rocks.

Meteorology

Meteorology is the study of the atmosphere. Meteorology is the study of the atmosphere, atmospheric phenomena, and atmospheric effects on our weather. The atmosphere is the gaseous layer of the physical environment that surrounds a planet. Earth's atmosphere is roughly 100 to 125 kilometers (65-75 miles) thick. Gravity keeps the atmosphere from expanding much farther.

Meteorology is a subdiscipline of the atmospheric sciences, a term that covers all studies of the atmosphere. A subdiscipline is a specialized field of study within a broader subject or discipline.

Climatology and aeronomy are also subdisciplines of the atmospheric sciences. Climatology focuses on how atmospheric changes define and alter the world's climates. Aeronomy is the study of the upper parts of the atmosphere, where unique chemical and physical processes occur. Meteorology focuses on the lower parts of the atmosphere, primarily the troposphere, where most weather takes place.

Meteorologists use scientific principles to observe, explain, and forecast our weather. They often focus on atmospheric research or operational weather forecasting. Research meteorologists cover several subdisciplines of meteorology to include: climate modeling, remote sensing, air quality, atmospheric physics, and climate change. They also research the relationship between the atmosphere and Earth's climates, oceans, and biological life.

Forecasters use that research, along with atmospheric data, to scientifically assess the current state of the atmosphere and make predictions of its future state. Atmospheric conditions both at the Earth's surface and above are measured from a variety of sources: weather stations, ships, buoys, aircraft, radar, weather balloons, and satellites. This data is transmitted to centers throughout the world that produce computer analyses of global weather. The analyses are passed on to national and regional weather centers, which feed this data into computers that model the future state of the atmosphere. This transfer of information demonstrates how weather and the study of it take place in multiple, interconnected ways.

Scales of Meteorology

Weather occurs at different scales of space and time. The four meteorological scales are: microscale, mesoscale, synoptic scale, and global scale. Meteorologists often focus on a specific scale in their work.

Microscale Meteorology

Microscale meteorology focuses on phenomena that range in size from a few centimeters to a few kilometers, and that have short life spans (less than a day). These phenomena affect very small geographic areas, and the temperatures and terrains of those areas.

Microscale meteorologists often study the processes that occur between soil, vegetation, and surface water near ground level. They measure the transfer of heat, gas, and liquid between these surfaces. Microscale meteorology often involves the study of chemistry.

Tracking air pollutants is an example of microscale meteorology. MIRAGE-Mexico is a collaboration between meteorologists in the United States and Mexico. The program studies the chemical and physical transformations of gases and aerosols in the pollution surrounding Mexico City. MIRAGE-Mexico uses observations from ground stations, aircraft, and satellites to track pollutants.

Mesoscale Meteorology

Mesoscale phenomena range in size from a few kilometers to roughly 1,000 kilometers (620 miles). Two important phenomena are mesoscale convective complexes (MCC) and mesoscale convective systems (MCS). Both are caused by convection, an important meteorological principle.

Convection is a process of circulation. Warmer, less-dense fluid rises, and colder, denser fluid sinks. The fluid that most meteorologists study is air. (Any substance that flows is considered a fluid.) Convection results in a transfer of energy, heat, and moisture—the basic building blocks of weather.

In both an MCC and MCS, a large area of air and moisture is warmed during the middle of the day—when the sun angle is at its highest. As this warm air mass rises into the colder atmosphere, it condenses into clouds, turning water vapor into precipitation.

An MCC is a single system of clouds that can reach the size of the state of Ohio and produce heavy rainfall and flooding. An MCS is a smaller cluster of thunderstorms that lasts for several hours. Both react to unique transfers of energy, heat, and moisture caused by convection.

The Deep Convective Clouds and Chemistry (DC_3) field campaign is a program that will study storms and thunderclouds in Colorado, Alabama, and Oklahoma. This project will consider how convection influences the formation and movement of storms, including the development of lightning. It will also study their impact on aircraft and flight patterns. The DC_3 program will use data gathered from research aircraft able to fly over the tops of storms.

Synoptic Scale Meteorology

Synoptic-scale phenomena cover an area of several hundred or even thousands of kilometers. High- and low-pressure systems seen on local weather forecasts, are synoptic in scale. Pressure, much like convection, is an important meteorological principle that is at the root of large-scale weather systems as diverse as hurricanes and bitter cold outbreaks.

Low-pressure systems occur where the atmospheric pressure at the surface of the Earth is less than its surrounding environment. Wind and moisture from areas with higher pressure seek low-pressure systems. This movement, in conjunction with the Coriolis force and friction, causes the system to rotate counter-clockwise in the Northern Hemisphere and clockwise in the Southern Hemisphere, creating a cyclone. Cyclones have a tendency for upward vertical motion. This allows moist air from the surrounding area to rise, expand and condense into water vapor, forming clouds. This movement of moisture and air causes the majority of our weather events.

Hurricanes are a result of low-pressure systems (cyclones) developing over tropical waters in the Western Hemisphere. The system sucks up massive amounts of warm moisture from the sea, causing convection to take place, which in turn causes wind speeds to increase and pressure to fall. When these winds reach speeds over 119 kilometers per hour (74 miles per hour), the cyclone is classified as a hurricane.

Hurricanes can be one of the most devastating natural disasters in the Western Hemisphere. The National Hurricane Center, in Miami, Florida, regularly issues forecasts and reports on all tropical weather systems. During hurricane season, hurricane specialists issue forecasts and warnings for every tropical storm in the western tropical Atlantic and eastern tropical Pacific. Businesses and government officials from the United States, the Caribbean, Central America, and South America rely on forecasts from the National Hurricane Center.

High-pressure systems occur where the atmospheric pressure at the surface of the Earth is greater than its surrounding environment. This pressure has a tendency for downward vertical motion, allowing for dry air and clear skies.

Extremely cold temperatures are a result of high-pressure systems that develop over the Arctic and move over the Northern Hemisphere. Arctic air is very cold because it develops over ice and snow-covered ground. This cold air is so dense that it pushes against Earth's surface with extreme pressure, preventing any moisture or heat from staying within the system.

Meteorologists have identified many semi-permanent areas of high-pressure. The Azores high, for instance, is a relatively stable region of high pressure around the Azores, an archipelago in the mid-Atlantic Ocean. The Azores high is responsible for arid temperatures of the Mediterranean basin, as well as summer heat waves in Western Europe.

Global Scale Meteorology

Global scale phenomena are weather patterns related to the transport of heat, wind, and moisture from the tropics to the poles. An important pattern is global atmospheric circulation, the large-scale movement of air that helps distribute thermal energy (heat) across the surface of the Earth.

Global atmospheric circulation is the fairly constant movement of winds across the globe. Winds develop as air masses move from areas of high pressure to areas of low pressure. Global atmospheric circulation is largely driven by Hadley cells. Hadley cells arc tropical and equatorial convection patterns. Convection drives warm air high in the atmosphere, while cool, dense air pushes lower in a constant loop. Each loop is a Hadley cell.

Hadley cells determine the flow of trade winds, which meteorologists forecast. Businesses, especially those exporting products across oceans, pay close attention to the strength of trade winds because they help ships travel faster. Westerlies are winds that blow from the west in the midlatitudes. Closer to the Equator, trade winds blow from the northeast (north of the Equator) and the southeast (south of the Equator).

Meteorologists study long-term climate patterns that disrupt global atmospheric circulation. Meteorologists discovered the pattern of El Nino, for instance. El Niño involves ocean currents and trade winds across the Pacific Ocean. El Niño occurs roughly every five years, disrupting global atmospheric circulation and affecting local weather and economies from Australia to Peru.

El Niño is linked with changes in air pressure in the Pacific Ocean known as the Southern Oscillation. Air pressure drops over the eastern Pacific, near the coast of the Americas, while air pressure rises over the western Pacific, near the coasts of Australia and Indonesia. Trade winds weaken. Eastern Pacific nations experience extreme rainfall. Warm ocean currents reduce fish stocks, which depend on nutrient-rich upwelling of cold water to thrive. Western Pacific nations experience drought, devastating agricultural production.

Understanding the meteorological processes of El Niño helps farmers, fishers, and coastal residents prepare for the climate pattern.

Meteorology Today

Today's meteorologists have a variety of tools that help them examine, describe, model, and predict weather systems. These technologies are being applied at different meteorological scales, improving forecast accuracy and efficiency.

Radar is an important remote sensing technology used in forecasting. A radar dish is an active sensor in that it sends out radio waves that bounce off particles in the atmosphere and return to the dish. A computer processes these pulses and determines the horizontal dimension of clouds and precipitation, and the speed and direction in which these clouds are moving.

A new technology, known as dual-polarization radar, transmits both horizontal and vertical radio wave pulses. With this additional pulse, dual-polarization radar is better able to estimate precipitation. It is also better able to differentiate types of precipitation—rain, snow, sleet, or hail. Dual-polarization radar will greatly improve flash-flood and winter-weather forecasts.

Tornado research is another important component of meteorology. Starting in 2009, the National Oceanic and Atmospheric Administration (NOAA) and the National Science Foundation conducted the largest tornado research project in history, known as VORTEX2. The VORTEX2 team, consisting of about 200 people and more than 80 weather instruments, traveled more than 16,000 kilometers (10,000 miles) across the Great Plains of the United States to collect data on how, when, and why tornadoes form. The team made history by collecting extremely detailed data before, during, and after a specific tornado. This tornado is the most intensely examined in history and will provide key insights into tornado dynamics.

Satellites are extremely important to our understanding of global scale weather phenomena. The National Aeronautics and Space Administration (NASA) and NOAA operate three Geostationary Operational Environmental Satellites (GOES) that provide weather observations for more than 50 percent of the Earth's surface.

GOES-15, launched in 2010, includes a solar X-ray imager that monitors the sun's X-rays for the early detection of solar phenomena, such as solar flares. Solar flares can affect military and commercial satellite communications around the globe. A highly accurate imager produces visible and infrared images of Earth's surface, oceans, cloud cover, and severe storm developments. Infrared imagery detects the movement and transfer of heat, improving our understanding of the global energy balance and processes such as global warming, convection, and severe weather.

Hydrometeorology

Hydrometeorology is a branch of meteorology and hydrology that studies the transfer of water and energy between the land surface and the lower atmosphere. Hydrologists often utilize meteorologists and data provided by meteorologists. As an example, a meteorologist might forecast 2–3 inches (51–76 mm) of rain in a specific area, and a hydrologist might then forecast what the specific impact of that rain would be on the local terrain.

UNESCO has several programmes and activities in place that deal with the study of natural hazards of hydrometeorological origin and the mitigation of their effects. Among these hazards are the results of natural processes and atmospheric, hydrological, or oceanographic phenomena such as floods, tropical cyclones, drought, and desertification. Many countries have established an operational hydrometeorological capability to assist with forecasting, warning, and informing the public of these developing hazards.

Hydrometeorological Forecasting

One of the more significant aspects of hydrometeorology involves predictions about and attempts to mitigate the effects of high precipitation events. There are three primary ways to model meteorological phenomena in weather forecasting, including nowcasting, numerical weather prediction, and statistical techniques. Nowcasting is good for predicting events a few hours out, utilizing observations and live radar data to combine them with numerical weather prediction models. The primary technique used to forecast weather, numerical weather prediction uses mathematical models to account for the atmosphere, ocean, and many other variables when producing forecasts. These forecasts are generally used to predict events days or weeks out. Finally, statistical techniques utilize regressions and other statistical methods to create long-term projections that go out weeks and months at a time. These models allow scientists to visualize how a multitude of different variables interact with one another, and they illustrate one grand picture of how the Earth's climate interacts with itself.

Risk Assessment

A major component of hydrometeorology is mitigating the risk associated with flooding and other hydrological threats. First, there has to be knowledge of the possible hydrological threats that are expected within a specific region. After analyzing the possible threats, warning systems are put in place to quickly alert people and communicate to them the identity and magnitude of the threat. Many nations have their own specific regional hydrometeorological centers that communicate threats to the public. Finally, there must be proper response protocols in place to protect the public during a dangerous event.

Climatology

Climatology is the branch of the atmospheric sciences concerned with both the description of climate and the analysis of the causes of climatic differences and changes and their practical consequences.

Earth – Sun Relationship

Generally the manner of the division of the year into seasons varies with latitude. In middle latitudes, the year is divided into 'autumn', 'winter', 'spring' and 'summer'. The terms 'summer' and 'winter' are not so significant in tropics, rather division into seasons is usually made in terms of rainfall amount as 'rainy season' and 'dry season' or in terms of the associated wind direction into 'south-west monsoon' and 'north-east monsoon' as in India. In the continental subtropical regions the natural seasons are usually defines in terms of temperature (hot or cold), or rainfall (rainy and dry), or both. In polar regions, the transition from summer to winter and vice versa is so sudden that spring and autumn largely disappear. Solar radiation is one of many sources of energy, and probably one of the most important sources, that drive environmental processes acting at the surface of the Earth. The amount and intensity of solar radiation reaching the Earth is affected by the geometric relationship of the Earth with respect to the Sun. The variations in the amount and intensity of solar radiation reaching the earth are affected by latitude, the rotation of the earth and its revolution around the sun. The study of the geometric relationship of the earth with respect to the sun explains why we have seasons.

Earth Rotation and Revolution

The term earth rotation refers to the spinning of the earth on its axis passing through the north and south poles. Turning in an eastward direction the earth rotates at a uniform rate once every 24 hours approximately, which is called a mean solar day. The orbit of the earth around the sun is called earth revolution. The earth's orbit around the sun is not circular, but elliptical with sun at its focus. An elliptical orbit causes the earth's distance from the sun to vary annually. The average distance of the earth from the sun is about 150 million kilometers. On 3rd January the earth is closest to the sun at an approximate distance of 147.5 million km, and this position is called as perihelion. On 4th July the earth is farthest from the sun at an approximate distance of 152.5 million km, and this position is called as aphelion. These annual variations in the earth-sun distance influence a slight change in the receipt of solar radiation. The difference in distance is not the cause of different seasons. Instead, seasons are caused by the tilt of earth's axis of rotation.

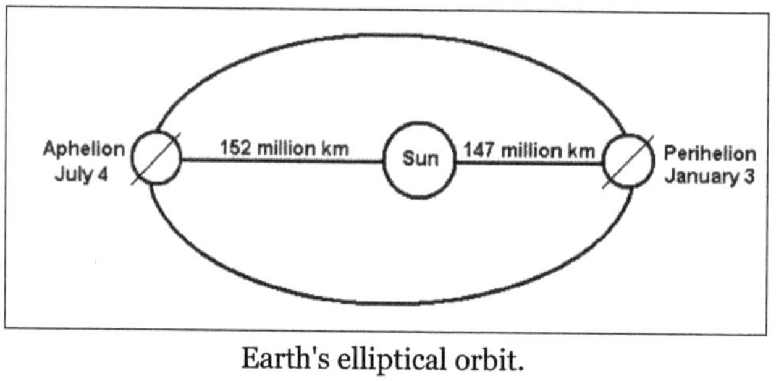

Earth's elliptical orbit.

Tilt of the Earth's Axis

The plane of the ecliptic is the plane of the Earth's orbit around the sun. The Earth's axis is tilted by $23\frac{1}{2}^{\circ}$ from the perpendicular to the plane of the ecliptic. In other words it makes an angle of $23\frac{1}{2}^{\circ}$ with the plane of elliptic as shown in figure. The axis of rotation remains pointing in the same direction as it revolves around the Sun. As a result, the earth's axis of rotation remains parallel to its position at any other time as it orbits the sun, a property called parallelism of axes.

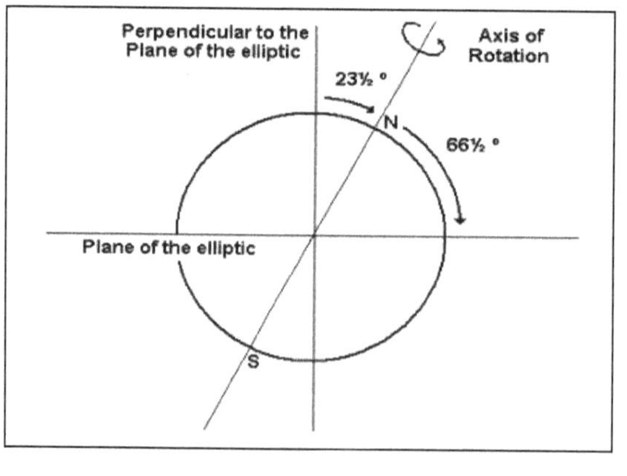

The constant tilt and parallelism causes changes in the angle that a solar rays makes with respect to a point on earth during the year, called the "sun angle". The most intense incoming solar radiation occurs where the sun's rays strike the earth at the highest angle. As the sun angle decreases, the beam of light is spread over a larger area and decreases in intensity due to the thickness of the atmosphere, increase in reflection and scattering of light.

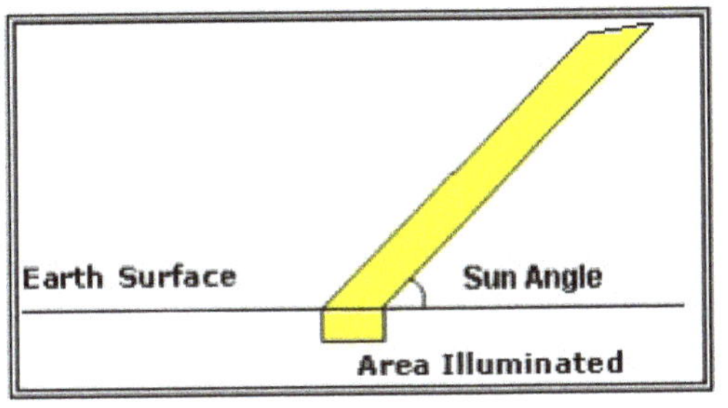

Solstices

On June 21 or 22, the axis of rotation of the earth is inclined towards the sun. The subsolar point, the place where the sun lies directly overhead at noon, is located at 23½ ° north latitude. This date is known as the summer solstice and marks the first day of the summer in the northern hemisphere.

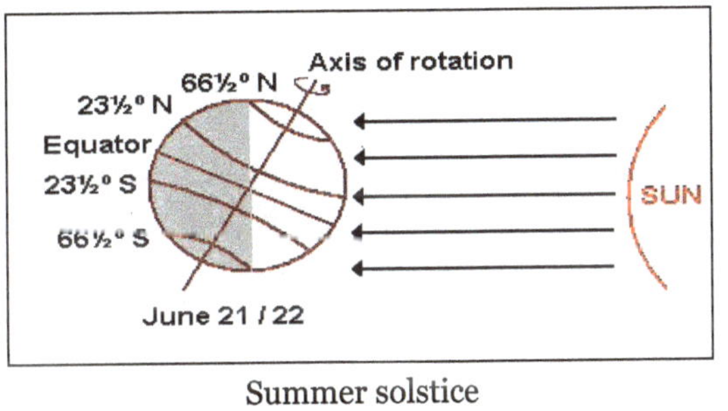

Summer solstice

This is the longest day of the year for places located north of Tropic of Cancer (23½° N latitude). As the noon's rays are vertical over 23½° N, the tangent rays in the northern hemisphere pass over the pole. This phenomenon keeps all places north of latitude of 66½° N in 24 hours of sunlight (polar day), while locations below latitude of 66½° S are in darkness (polar night). The winter solstice occurs on December 21 or 22 when the earth has oriented itself so the North Pole is facing away from, and the South Pole into the Sun. The Sun lays directly overhead at noon at 23½° S latitude, called as Tropic of Capricorn. The places poleward of 66½° S latitude receives 24 hours of daylight and the places poleward of 66½° N are in the darkness. The winter solstice refers to the first day of winter in the northern hemisphere.

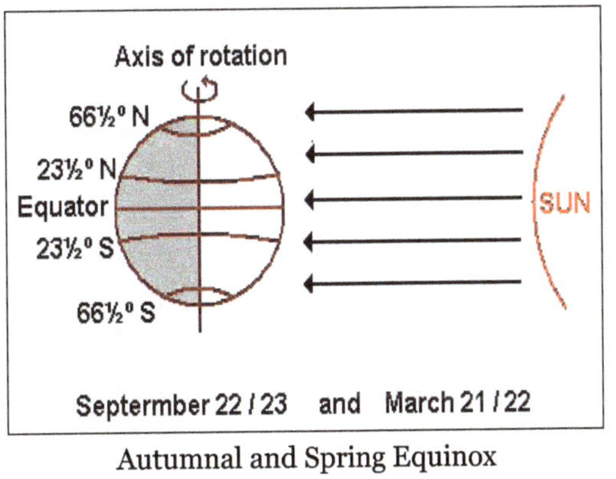

Autumnal and Spring Equinox

Equinoxes

Midway between the solstices are two dates when the sun shines directly on the equator. The axis of rotation is still inclined but it is tilted sideways with respect to the sun rather than towards or away from the sun. at these times, the tangent rays strike the poles so that the days and nights are equal over the entire earth. The period between summer and winter is called as autumn. The autumnal equinox on September 22 or 23 indicates the beginning of autumn season in the northern hemisphere. March 21 or 22 is the first day of the spring season and as such this date is called as the spring equinox. Equinoxes mark the seasons of autumn and spring and are a transition between the two more extreme seasons, summer and winter.

The Seasons

During the summer months the earth is inclined toward the Sun yielding high sun angles whereas during the winter, the earth is oriented away from the Sun creating low sun angles. The tilt of the earth and its impact on sun angle is the reason the northern and southern hemispheres have opposite seasons. Summer occurs when a hemisphere is tipped toward the Sun and winter when it is tipped away from the Sun.

Annual March of Seasons

As seen in figure, over the course of a year, the sun's rays are perpendicular to the surface (directly overhead) at places between Tropic of Cancer $23\frac{1}{2}^{0}$ N and Tropic of Capricorn $23\frac{1}{2}^{0}$ S latitudes only. Places between the two tropics experience two times when the sun is directly overhead over the course of a year and the sun angle does not vary much over these places. Poleward of this region there is greater variation in the sun angle and consequently the greater variation in surface heating.

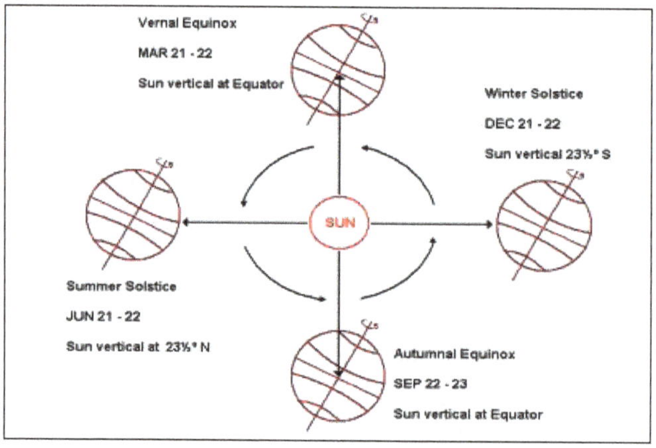

Annual March of Seasons

Solar Radiation

The weather results from the interaction of solar radiation on the earth's atmosphere and surface. The two movements (viz. rotation and revolution) explain the changing elevation of the sun as well as the latitudinal and seasonal variations in length of the day, receipt, and escape of radiation and weather. The sun at a temperature of about 6000 K^{0} is the source of nearly all of our energy. The earth intercepts an infinitesimally small part of the sun's output, 5×10^{-10} %. Only a portion of the sun's radiation reaches the earth's surface as direct radiation, the remainder being reflected, absorbed, or scattered by the atmosphere. The maximum emission of solar radiation occurs at relatively short wavelengths in the visible spectrum (between 0.4 μm and 0.7 μm). The sun radiates approximately $5^6 \times 10^{26}$ cal of energy per minute. The energy per unit area incident on the earth is equal to:

$$S = \frac{56 \times 10^{26} \, \text{cal} \, \text{min}^{-1}}{4\pi (1.5 \times 10^{13} \, \text{cm})^2} \approx 2.0 \, \text{cal} \, \text{cm}^{-2} \, \text{min}^{-1}$$

S is called as the Solar Constant. {1.5 cm 10^{13} × is the mean distance of the earth from the sun}

The Solar Constant is defined as the flux of solar radiation at the outer boundary of the earth's atmosphere that is received on a surface held perpendicular to the sun's direction at the mean distance between the sun and the earth. The energy per unit area is expressed in Langleys (ly) and 1 ly = 1 cal cm^{-2}. Thus the solar constant for the earth is 2 Ly.

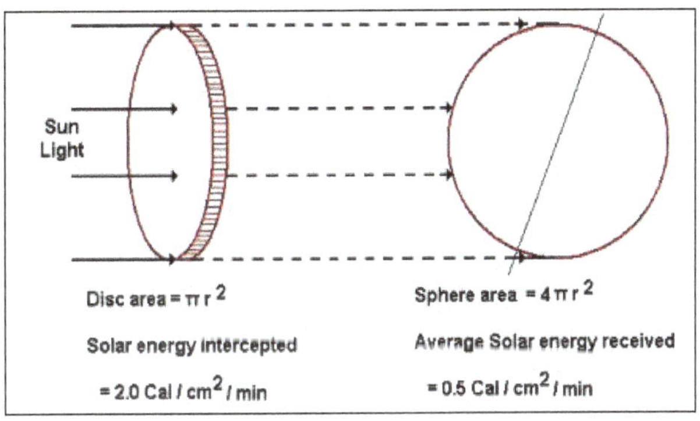

If the surface, oriented perpendicular to the sun's rays, is thought of as a circular non-rotating disc with a radius r equal to the radius of the earth, the sun facing side of the disc will intercept the same amount of solar radiation as does the spherically shaped rotating earth.

Since the area of a sphere is four times the area of one side of a disc ($4\pi r^2$ versus πr^2), as shown in figure, the global average amount of energy received at the top of the atmosphere is 0.5 ly min^{-1}, or one quarter of the solar constant.

Terrestrial Radiation

The absorption of solar radiation raises the temperature of the earth's surface and its atmosphere. The radiation emitted from the land and water surface of the earth is long-wave radiation, in contrast to the short-wave radiation it receives from the sun. The earth approaches an average temperature of 294 K°, and gives out long-wave radiation in the range 4μ-80μ with a maximum at 10μ. Water vapour in the atmosphere is a strong absorber of long waves, particularly between 5.5μ and 7μ and above 27μ.

Albedo

It is defined as the ratio of the reflected radiation to the total intercepted radiation. It is described in terms of percentage of reflected radiation. The albedo of the earth-atmosphere system is 0.30. The moon has an albedo of only about 0.07, indicating that it absorbs most of the solar radiation striking its surface. Thus, viewed from space, the earth shines more brilliantly than the moon. The major reason for this is the presence of clouds. The moon has no atmosphere and no cloud.

Some typical albedo are:

Fresh snow	0.75 – 0.90
Cloud tops	0.60 – 0.90
Old snow	0.50 – 0.70
Sand	0.15 – 0.35
Seas (high sun angle)	0.05 – 0.10
Forests	0.03 – 0.10

Terrestrial Heat Balance

The mean temperature of the earth undergoes fluctuations of varying periods. Averaging over a period of number of years can smooth the more rapid fluctuations out. The remaining slower variations provide evidence of a net gain or loss of energy during the period of averaging. The changes occur so slowly, however, that the net energy change per year is very small. In general, there seems to be no over all significant trend in the mean temperature of the earth or its atmosphere during the interglacial periods. Thus a long-term mean heat balance exists at each point of the earth and its atmosphere.

The total amount of solar radiation received on a horizontal surface is about 0.5 ly min^{-1}. In this explanation, the figures represent estimates for the northern hemisphere but the difference between the northern and southern hemisphere are presumed to be not large.

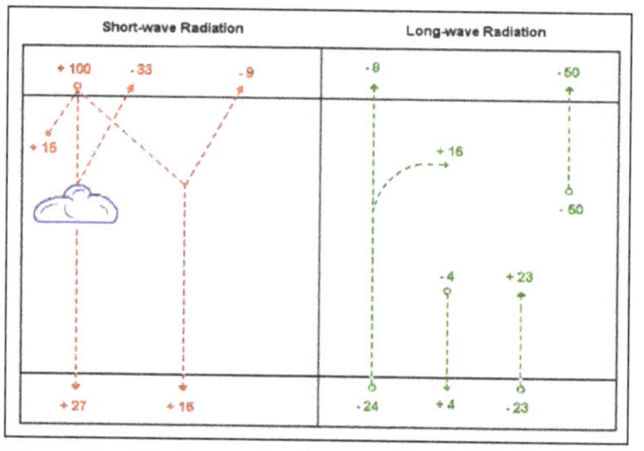

Mean heat balance in the northern hemisphere

Out of these 27 % penetrates directly to the earth's surface and 16 % arrives as the diffuse sky radiation. So that 43 % reaches the ground together. The atmosphere including clouds absorbs 15 %. The remaining 42 % is reflected back into space, which represents the albedo of the earthatmosphere system. It is composed of the reflection on clouds and on the ground, which accounts for 33 % and the diffuse reflection, which makes up the remaining 9 %.

It is noticed that the diffuse radiation towards the ground (16 %) is considerably greater than that returned to space (9 %). This difference is due to the fact that the larger dust particles scatter more radiation in the direction away from the sun than in the direction towards the sun.

In the figure, the radiation received by the earth and the atmosphere is counted positive and the radiation emitted or reflected and scattered to space is denoted by negative.

The 42 % of the incoming solar radiation is returned directly back to the space and the remaining 58 % is absorbed by the ground and the atmosphere. This 58 % must be radiated back to space since the yearly mean temperature of the earth as a whole remains the same. The radiation from the ground upwards is called the 'effective radiation'.

The 24 % represents the difference between the actual radiation from the ground and the radiation from the atmosphere to the ground. Out of this 24 %, 16 % is reabsorbed in the atmosphere while 8 % returns directly to the space. The other 50 % is radiated back to space by the atmosphere.

When the heat transports by turbulence and by condensation are taken into consideration, a heat balance exists also for the earth's surface and the atmosphere separately. This transport of heat is estimated at 4 % and 23 % respectively.

The following is the summary of the separate heat balance of the earth:

The surface of the earth			
Receives		Loses	
By direct radiation	27 %	By radiation	24 %
By diffuse radiation	16 %	By condensation (evaporation)	23 %
By turbulent transfer	4 %		
Total =	47 %	Total=	47 %

The atmosphere			
Receives		Loses	
By absorption of solar radiation	15 %	By radiation	50 %
By absorption of ground radiation	16 %	By turbulent transfer	4 %
By condensation	23 %		
Total =	54 %	Total=	54 %

Distribution of Solar Radiation

The unequal distribution of solar radiation over the earth is the primary cause of weather and climate. The rate of receipt of solar energy varies with latitudes, seasons and time of the day i.e. the angle of sun's rays with the surface of the earth.

A) Distribution of solar radiation without atmosphere:

The annual march of the solar radiation at the top of the atmosphere is shown figure.

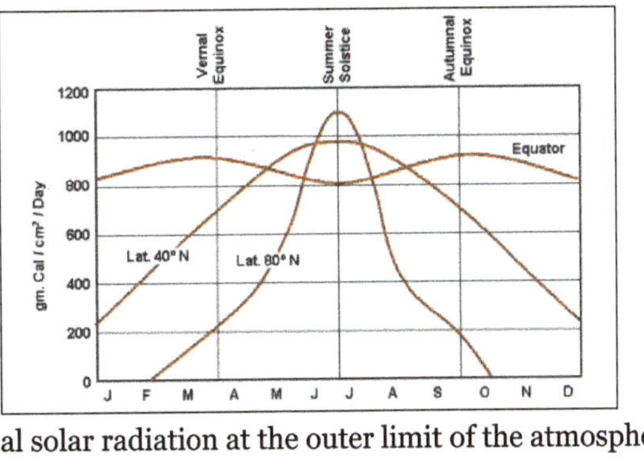

Annual solar radiation at the outer limit of the atmosphere.

In tropics, the intensity of solar radiation remains quite high throughout the year with little seasonal variation. The noon rays are vertical twice a year at all places situated between two tropics. As a result the solar radiation curve shows two maxima and two minima for the low latitudes.

The mid latitude curve (40°) is broadly representative of the belts lying between 23½° and 66½° in each hemisphere. It shows a single strong maximum and a single minimum, both of which coincide with the solstice. The curves show a large seasonal variation.

The high latitude (80°), which represents belts poleward of Arctic and Antarctic circles, resembles that of middle latitudes. The only difference in this curve is that it reaches zero during winter when there is no solar radiation.

B) Distribution of solar radiation with atmosphere (without clouds):

The distribution of solar radiation with atmosphere but without clouds is shown in figure.

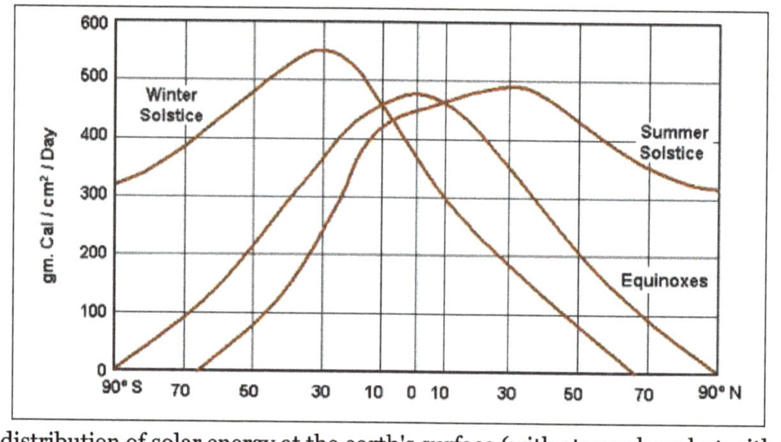

Latitudinal distribution of solar energy at the earth's surface (with atmosphere but without clouds).

A solar beam, while passing through the atmosphere (without clouds), is depleted by scattering, reflection and absorption. Thus the amount of radiation reaching the surface of the earth is less than that received at the outer limits of the atmosphere. The depletion is maximum at high latitudes due to higher obliquity of the solar beam resulting in the path of the beam passing through much greater thickness of the atmosphere than in the lower latitudes.

At the time of the equinoxes the latitudinal distribution of solar radiation is symmetrical about the equator with the amount decreasing to zero at each pole.

During summer solstice the more nearly vertical solar beam and larger days combined together produce a broad maximum in middle latitudes of northern hemisphere. The same feature is observed in the middle latitudes of southern hemisphere during the winter solstice.

Latitudinal variation of radiation is small in summer hemisphere compared to the winter hemisphere.

C) Distribution of solar radiation with atmosphere and clouds:

The clouds reflect a large amount of solar radiation reducing the amount of it reaching the surface of the earth. The maximum total annual radiation is not found at the equator but rather at about latitude 20°N and 20°S. the lesser amount near the equator in the southern hemisphere is due to greater cloudiness and more ocean surface in the southern hemisphere.

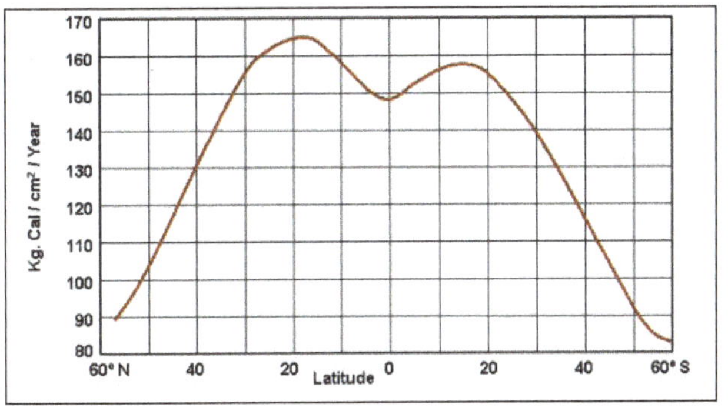

Total annual solar radiation received at the earth's surface by latitude belts.

In the equatorial belt, area of least solar radiation coincides with the warm continents, where convective clouds are in abundant. Maximum solar radiation is received in the sub-tropics, which are relatively less cloudy region. In the high latitudes, the lowest annual radiation values are over oceans because of abundant cloudiness.

Weather, Climate and Climatic Elements

Weather is the instantaneous state of the atmosphere, or the sequence of the states of the atmosphere as time passes. The difference between climate and weather is usefully summarized by the popular phrase "Climate is what you expect, weather is what you get". Weather and climate are meteorological terms that are related but not interchangeable.

Climate is usually defined as the average weather plus the extremes for a given time period and a given location. More technically, for a given location and time period the climate is the probability distribution of each variable that characterizes the local weather. Climate changes from one location to another, and in a given location climate can change from one time period to another. Weather is the sum total of the atmospheric variables at a given place for a brief period of time; it is an everyday experience. Thus we speak of today's weather, or of last week's. Climate, on the other hand, refers to a more enduring regime of the atmosphere.

According to The Meteorological Glossary, published by the UKMO, 'climate' is defined as 'the synthesis of the day-to-day values of the meteorological elements that affect its locality'. Synthesis here means more than simple averages as the climate also involves extreme values, and the ranges within which phenomena occur, and the frequencies of weather types with associated values of elements.

Climatic data are usually expressed in terms of an individual calendar month or season and are determined over a period of about 30 years. This is long enough to ensure that representative values for the month or season are obtained and freaks or abnormal values do not exert too strong an influence.

Elements of Climate and Weather

Although weather and climate are not identical both are described by combinations of the same atmospheric variables, called the elements of weather and climate. Primarily these element are Pressure, Temperature, Precipitation, Humidity, Wind and Cloudiness. The atmospheric pressure is of particular importance in determining the characteristics of the other variables. It is atmospheric pressure that to an important degree determines the direction and speed of the wind, and it is the wind that in turn moves air masses of contrasting temperature and moisture from one locality to another. While air movement is predominantly in a horizontal direction, there is also some slight upward or downward movement. Where the motion is upward, cloud and precipitation are likely, while downward air movement, or subsidence, favors fair skies.

Controls of Climate

Climatic Controls are the factors affecting the climate of particular place. The most fundamental control of both weather and climate is the unequal heating and cooling of the atmosphere in different parts of the earth. While the earth as a whole loses as much heat to space as it gains from the sun, some parts experience a net gain and others a net loss. The unequal heating occurs on a wide variety of geographic scales, the largest and most important of which is the differential between high and low latitudes. But heating and cooling differences also exist between continents and oceans, between snow-covered and snow-free areas, between forested and cultivated land, and even between cities and their surrounding country sides. These heating and cooling differences, and the air movements (winds) they induce, represent the overall general background control of weather and climate. The more specific controls are derived from various geographic factors.

A) Latitudinal Variations in Solar Radiation:

Latitudinal differences in the amounts of solar energy received are the most basic climatic control. In low latitudes the sun is high in the sky, the solar radiation is intense, and the climate is warm and tropical; in high latitudes the sun is lower in the sky, the solar radiation is weaker, and the climate is colder. The zone of maximum solar radiation shifts northward and southward during the year, thereby producing the seasons. The effectiveness of solar heating also varies with the nature of the surface on which the sunshine falls. Thus a strongly reflecting snow surface is heated much less than a land surface lacking snow.

B) Altitudes:

Since within the troposphere temperature normally decreases with increasing Altitude, places at higher elevations are likely to have lower temperatures (and often also more precipitation) than adjacent lowlands. Thus altitude is a climatic control. Where a high mountain chain lies athwart the path of prevailing winds, it acts to block the movement of air and hence the transfer of warm or cold air masses. In addition, the upward thrust of air on a mountain's windward side and the downward movement of air on its lee side tend to make for increased precipitation in the former instance and a decrease in the latter.

C) Distribution of Continents and Oceans:

Continents heat and cool more rapidly than do oceans. Consequently non coastal continental areas experience more intense summer heat and winter cold than do oceanic and coastal areas.

D) Pressure and Wind Systems:

Differences in heating and cooling between high and low latitudes, between land and water areas, and between snow-covered and bare land surfaces lead not only to regional temperature contrasts but also to differences in atmospheric pressure which in turn induce air movements (winds). Air in motion, which in itself is an important element of weather and climate, also operates as a control, for it serves as a transporter of heat from regions of net heat gain to regions of net loss. And just as there is a great variety of geographic scales of differential heating and cooling of terrestrial surfaces, so there is a great variety of scales relating to atmospheric pressure and atmospheric motion. They range from those of hemispheric magnitude, such as the belts of westerly winds in middle latitudes and the belts of easterlies that encircle the low latitudes, to the small but extremely violent tornado.

The mobile low-and high-pressure systems which bring day-to-day weather changes and are conspicuous features on daily weather maps are of a common scale of atmospheric motion. The frequency of occurrence and the paths followed by these transient mobile pressure and winds systems are important factors in determining climate. Some pressure and wind systems, especially the highs over the subtropical oceans, tend to be semi permanent in position, and they too are of great climatic importance.

E) Ocean Currents:

Ocean currents, both warm and cold, which are largely induced by the major wind systems, are also an important climatic control. They are highly important in transporting warmth and chill in a north-south direction, and in so doing give some coastal regions distinctive climates.

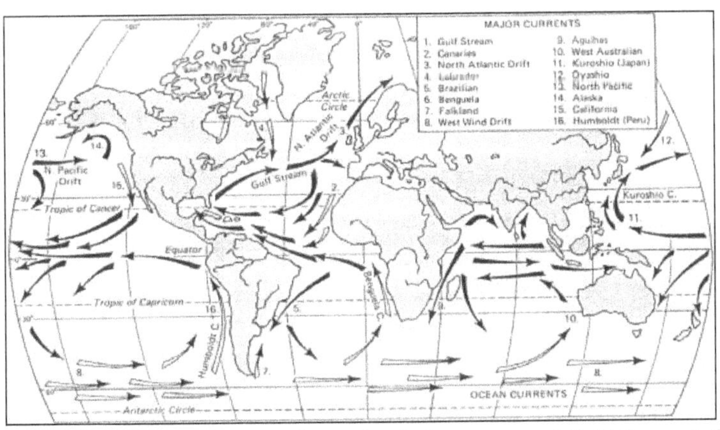

Major Ocean Currents. (Warm Currents are shown by dark arrows and Cold Currents by open arrows).

F) Local Features:

Finally, the climate of a place is affected by a variety of local features, such as its exposure, the slope of the land, and the characteristics of vegetation and soil. In the Northern Hemisphere south-facing slopes receive more direct sunlight and have a warmer climate than those with a northern exposure, which not only face away from the sun, but are also more open to cold northerly winds. Areas with sandy, loosely packed soil, because of their low heat conductivity, are inclined to experience more frosts than do areas with hard packed soils; valleys normally have more frequent and severe frosts than the adjacent slopes; and cities are usually warmer than the adjacent country sides.

General Circulation of the Atmosphere

The simplest observed global characteristic of the atmosphere is that the tropics are much warmer than the poles. As discussed earlier, this is a straightforward consequence of the geometry of the earth. The annually averaged incoming solar radiation per unit area of the earth's surface is much greater at the equator than at the poles. The difference arises because of the fact that the polar regions are covered in ice and snow and therefore reflect much of the incoming radiation back to space. Another fact is that the tropical regions actually receive more energy from the sun than they emit back to space, while the converse is true in high latitudes.

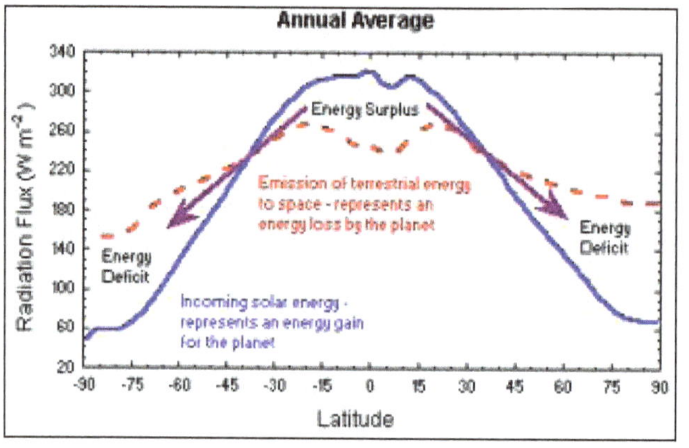

Latitudinal variation in annual average of radiation.

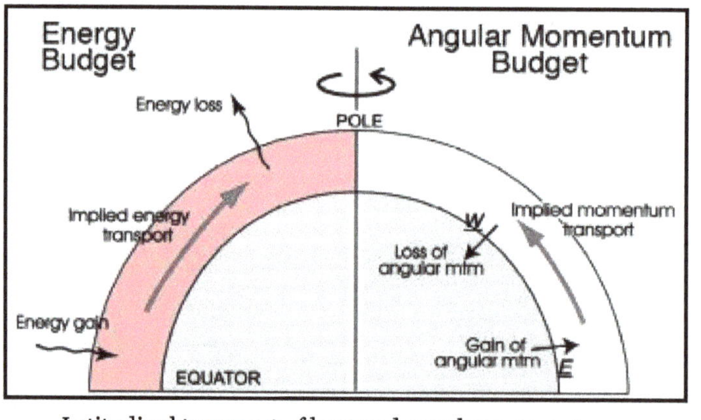

Latitudinal transport of heat and angular momentum.

Over the globe, the energy balance is nearly balanced when averaged over a year (incoming equals outgoing). Hence, there must be a process acting to transport excess energy from the tropics to make up the deficit in high latitudes, as depicted schematically in figure. To compensate for the surplus and deficit of radiation in different regions of the globe, atmospheric and oceanic transport processes distribute the energy equally around the earth. This transport is accomplished by atmospheric winds and ocean currents.

Single-Cell Model of the General Circulation

If the Earth's surface were smooth, uniform, and stationary, atmospheric circulation would be very simple. The atmosphere would act as a contained fluid and movements within this fluid would be the convective currents caused by temperature and density differences.

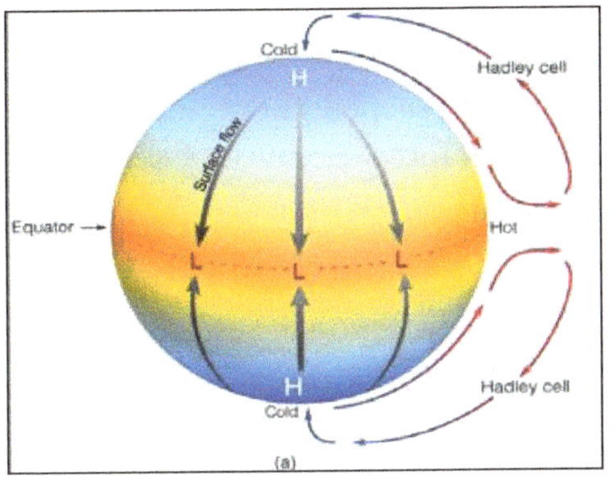

Single-Cell Model of General Circulation.

The latitudinal transfer of heat would result in a single circulation cell, where the surface air converges and rises at the equator, spreads laterally toward the poles, descends and flows back toward equator at the surface.

This creates the single cell circulation model possible only on a non-rotating Earth and sun being directly over the equator.

Three-Cell Model of the General Circulation

This model represents the average circulation of the atmosphere and is used to describe the atmospheric transport of energy. It considers effects of coriolis force due to the Earth's rotation. In this circulation model, the Northern and Southern Hemisphere are each divided into three cells of circulation, each spanning 30 degrees of latitude. The latitudes that mark the boundaries of these cells are the Equator, 30° North and South, and 60° North and South.

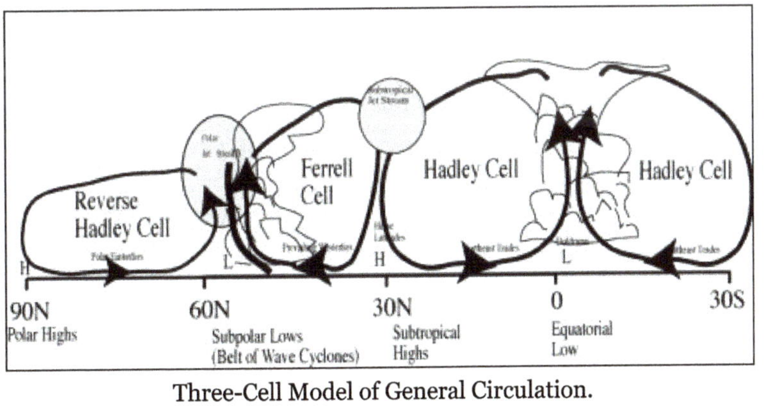

Three-Cell Model of General Circulation.

Hadley Cell

George Hadley, an English meteorologist, theorized this first circulation cell in 1735. The Hadley cell is the strongest of the three cells of circulation and is formed as warm air rises above the Equator and starts to flow northward.

The rising air cools condenses and forms a region of intense clouds and heavy precipitation. This area is called the Inter-Tropical Convergence Zone (ITCZ) and corresponds regions over which the tropical rain forests are found. The ITCZ moves north and south following the sun during the year. Because the stratosphere is stable, rising air that reaches the tropopause, moves poleward. By the time the air moving northward reached about 30° N it has become a westerly wind (it is moving to the east) due to the Coriolis force. Because of conservation of angular momentum, the poleward moving air increases speed. The increased speed and the Coriolis force are responsible for the subtropical jet.

This poleward moving air leads to the formation of semi-permanent high pressure belt at the surface that results from the sinking air at 30°.

Once the sinking air reaches the ground, some flows to the equator, turning west (in the northern hemisphere) as it goes due to the Coriolis force. This surface air forms the trade winds which blow steadily from the northeast in the northern hemisphere and southeast in the southern hemisphere.

Polar Cell

This is the northernmost cell of circulation and its mean position is between 60°N and the North Pole. At the pole, cold, dense air descends, causing an area of subsidence and high pressure. As the air sinks, it begins spreading southward. Since the coriolis force is strongest at the poles, the southward moving air deflects sharply to the right. This wind regime is called the surface polar easterlies, although the upper winds are still predominantly from the southwest. Near 60°N, the southeasterly moving air moving along the surface collides with the weak, northwesterly surface flow that resulted from spreading air at 30°N. This colliding air rises, creating a belt of low pressure near 60°N.

Ferrel Cell

The mid-latitude circulation cell between the Polar cell and the Hadley cell is called the Ferrel cell. The Ferrel cell circulation is not as easily explained as the Hadley and Polar cells. Unlike the other two cells, where the upper and low-level flows are reversed, a generally westerly flow dominates the Ferrel cell at the surface and aloft. It is believed the cell is a forced phenomena, induced by interaction between the other two cells. The stronger downward vertical motion and surface convergence at 30°N coupled with surface convergence and net upward vertical motion at 60°N induces the circulation of the Ferrel cell. This net circulation pattern is greatly upset by the exchange of polar air moving southward and tropical air moving northward.

Distribution of Wind and Pressure Over the Surface of the Earth

In subtropical region, the northeast trade winds and southeast trade winds prevail in the northern and southern hemisphere respectively. They are separated near the equator by a zone of prevailing calms known as 'doldrums'. However, the doldrums are frequently absent in one part or the other of the equatorial zone. Then the northeast trades and southeast trades meet directly at a surface of discontinuity on which strong rainsqualls may develop. The zone of doldrums is largely north of equator. Over some parts of the earth it even extends into southern hemisphere. Where the zone of doldrums is displaced far north from its symmetrical position at the equator, the southeast trade winds may reach into the northern hemisphere and get deflected into southwest winds under the influence of the coriolis force.

Beyond the trade wind regions of each hemisphere are the zones of 'prevailing westerlies'. These westerlies are better developed in the southern hemisphere than in the northern hemisphere. These zones of westerlies are the regions of the migrating cyclones and anticyclones.

Poleward from the region of westerlies, beyond 65 to 75 latitude easterly winds with a component towards the equator prevail.

Near the equator, a zone of low pressure is found. It is followed in both hemispheres by zone of high pressure called 'subtropical' high-pressure belts. In summer hemisphere, these high-pressure belts do not extend continually around the earth but are interrupted over the continents since the land is hotter than the oceans. The central regions of the high-pressure belts have frequent calms.

In the region of the westerlies even the mean pressure distribution is rather irregular because of the migrating pressure centers. On the whole, the pressure decreases here toward the pole.

In the surface layers of the arctic and of the Antarctic zone, the pressure distribution is anticyclonic, because of the low temperature of the air in the surface layers. Since Greenland and Antarctica are not symmetrical with respect to the poles, the zones of the easterly winds in the polar regions are not symmetrical to the poles.

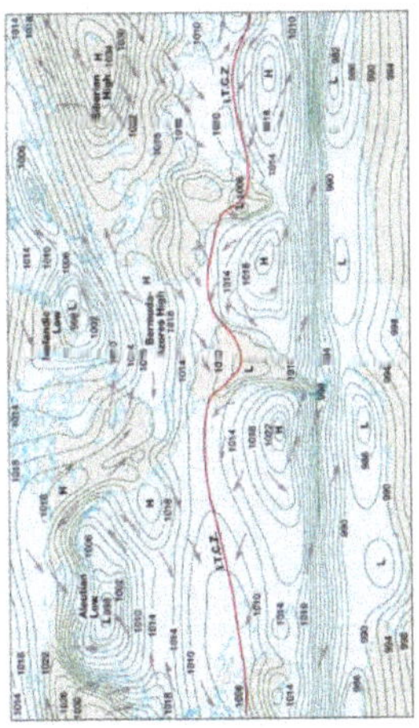

Mean Sea Level Pressure (hPa) in January.

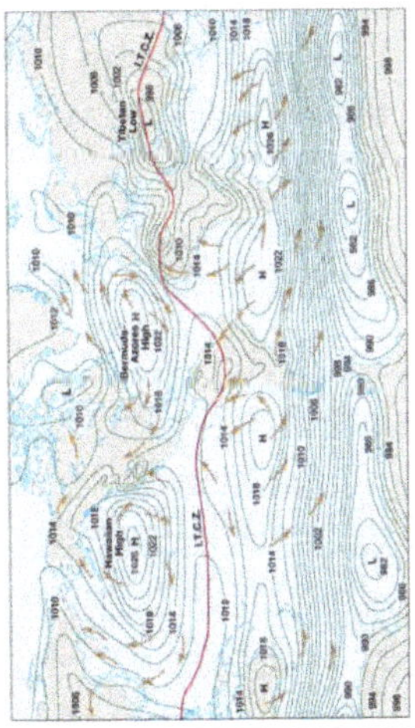

Mean Sea Level Pressure (hPa) in July.

Effect of Land and Sea on the Wind and Pressure Distribution

The unequal heating of water and land produces considerable modifications of the pressure and wind distribution. These deviations are more noticeable during the extreme months of January and July.

During July, the large continental mass of the northern hemisphere is much warmer than the surrounding oceans. Hence a low pressure is found over the continents, which interrupt the high-pressure belt. The continental low is in direct connection with the equatorial low-pressure zone of doldrums. The southern hemisphere has winter in July so that here the land is colder than the water. But the disturbance of the pressure distribution is much less because of the smaller land area in the southern hemisphere.

Over the continents, in the northern hemisphere during July, the northeast trade winds are absent. Instead, the southwest winds that are observed are continuation of the southeast trade winds of the southern hemisphere, which are deflected to southwest as they cross the equator. The deflection of the southeast trades into the southwest monsoon is especially pronounced over India and the Indian Ocean.

During January in the northern hemisphere, the continent is much colder than the adjoining oceans where a strong high is developed. On the south side of the continental anticyclone, the winds are northeasterly. The centers of low pressure over the oceans are considerably intensified during the winter so that the cyclonic activity is now at a maximum. On the north side of the continental anticyclone, the winds are westerly. Thus the belts of westerlies surround the northern hemisphere completely, but are displaced more northward over the continents.

The southern hemisphere has its summer during January. Thus the pressure over the continents is now lower than over the adjoining oceans. As a result, the zone of doldrums with its low pressure shows a southward bulge over the continent. But the interruption of the southern subtropical high-pressure belt by the continents during January is only slight.

Distribution of Temperature Over the Surface of the Earth

The atmospheric temperature is controlled principally by the incoming soalr radiation. Hence its distribution depends largely on the latitude. Also, it is affected by the nature of the surface of the earth especially by the differences between water and land, by the altitude, and by the prevailing winds.

Considering the mean temperature of each latitude, the latitudinal control of temperature is found more dominant.

Table: Latitudinal distribution of temperature (°F).

Latitude	Year	January	July	Range
90° N	-8.9	-42.0	30.0	72.0
80° N	-1.0	-26.0	35.6	61.6
70° N	12.7	-15.3	45.1	60.4
60° N	30.0	3.0	57.4	54.1
50° N	42.5	19.2	64.6	45.6
40° N	57.4	41.0	75.2	34.2
30° N	68.7	58.1	81.1	23.0
20° N	77.5	71.2	81.1	23.0
10° N	80.1	78.4	80.4	2.5
Equator	79.2	79.5	78.1	1.8
10° S	77.5	79.3	75.0	4.3
20° S	73.2	77.7	68.0	9.7
30° S	61.9	71.7	58.5	12.9
40° S	53.4	60.4	48.2	11.9
50° S	42.4	46.6	38.1	8.5

60° S	25.9	35.8	15.6	20.2
70° S	7.5	25.7	-9.4	35.1
80° S	-16.6	12.6	-39.1	51.7
90° S	-27.6	7.7	-54.0	62.0

The yearly averages show that the warmest latitude is not at the equator but at 10° N. Equator is the hottest circle of latitude only during January. In the month of July, the warmest parallel latitude is somewhat north of 20° N.

The northern hemisphere as a whole has a higher annual mean temperature than the southern hemisphere. Also on the yearly average, each latitude of the northern hemisphere is warmer than the corresponding latitude in the southern hemisphere. The temperature decrease with latitude is very small in tropical and sub-tropical latitudes.

During January the northern hemisphere has its winter. The temperature decrease from the equator toward either pole is considerably greater in winter than in summer (due to the seasonal variation of radiation). The greatest surplus of radiation during January occurs at about 30° south, while the temperature maxima is found at the equator. This is because of the fact that the temperature does not depend on the radiative balance alone.

The meridional temperature gradient acts as a regulator of the intensity of the general circulation; hence the general circulation is considerably more intense during the winter season than during the summer season.

The annual range of temperature is the difference between the mean temperature of the warmest and of the coldest month. It is notices that at the equator and at 10° N latitude the range is not equal to the difference between the mean temperatures of July and January because of the reason that at these latitudes the monthly maxima and minima are not reached in July and in January.

The annual temperature range is very small in tropical latitudes and increases towards both poles since the yearly variation of incoming solar radiation is larger at poles than at the equator. South of 30° S latitude the annual temperature range decreases to a secondary minimum at 50° south. This drop is due to the sharp decrease of continental area beyond 30° S latitude.

Table: Shows the mean temperature during January and july and during the whole year for each hemisphere separately and for the whole earth.

Mean temperatures (°F)				
Part of the globe	Year	January	July	Annual Range
Northern hemisphere	59.3	46.5	72.3	25.8
Southern hemisphere	56.0	62.7	49.5	13.2
Whole earth	57.7	54.6	60.8	6.2

It is seen that northern hemisphere has higher annual mean temperature and large annual range than the southern hemisphere. The higher annual range is due to the much larger continental areas in the northern hemisphere. Land surfaces are heated much more strongly in summer and cooled to much lower temperatures in winter than the water surfaces.

The greater land covers in the northern hemisphere is not the only factor responsible for its higher mean temperature. The oceans are also warmer in the northern than in the southern hemisphere. This is due to:

a) The transport of warm equatorial waters from the southern hemisphere into the northern hemisphere by the southeast trades, which cross the equator.

b) The partial protection of its oceans from the cold polar waters and from the Arctic ice by land barriers, there being no such barrier between the Antarctic Ocean and the southern oceans in more equatorial latitudes.

Effect of Land and Sea on the Temperature Distribution

The effect of land and water on the temperature distribution can be seen in figures, which show the sea-level isotherm during the months of January and July.

As seen in figure, over large part of the globe the January isotherms are not parallel to the latitude circles. In the northern hemisphere, the isotherms are deflected toward lower latitudes over the continents, indicating that the continents are colder than the oceans. The coldest region is found in Siberia.

During January a sweep of isotherms over eastern part of the north Atlantic is seen. This is due to the combination of the effects of north Atlantic drift and the prevailing westerly winds.

The prevailing winds everywhere in temperate latitudes are westerly. Consequently, during the winter the west coasts of the temperate zones are warmer than the east coasts.

In the southern hemisphere, January is the summer month when the continents are warmer than the oceans.

On the whole the isotherms in the southern hemisphere follows the parallels of latitude more closely than in the northern hemisphere since the earth's surface is more uniform.

During July, the continents of the northern hemisphere are warmer than the adjoining oceans. Over north Africa, the Arabian peninsula and southwestern Asia, regions surrounded by closed 90 °F isotherms is found. In the southern hemisphere, which now has its winter, the isotherms again have a much more regular shape than the isotherms over the northern hemisphere owing to the smaller continental areas.

The meridional temperature gradient is much greater during winter than during the summer.

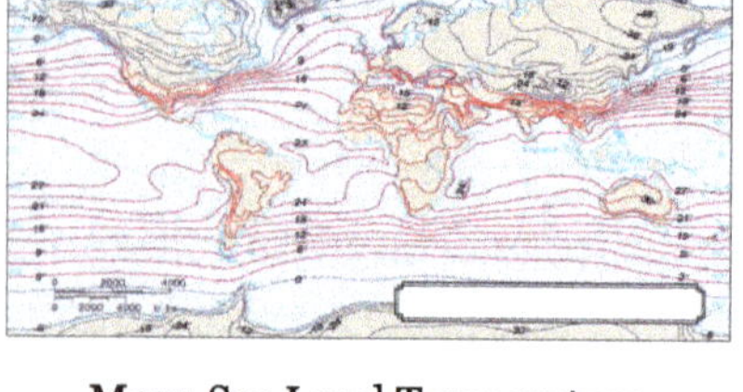

Mean Sea Level Temperature (°C) in January.

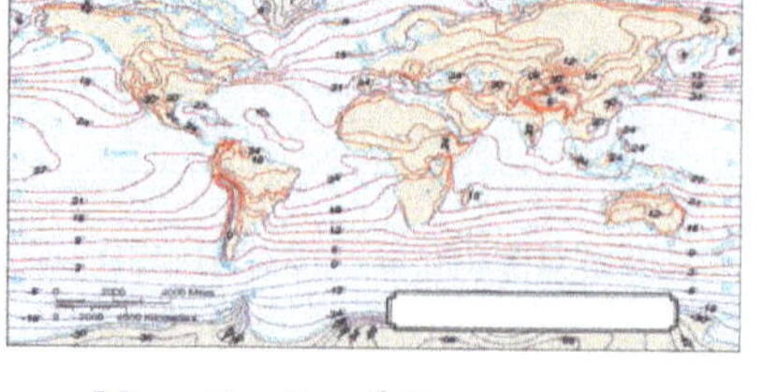

Mean Sea Level Temperature (°C) in July.

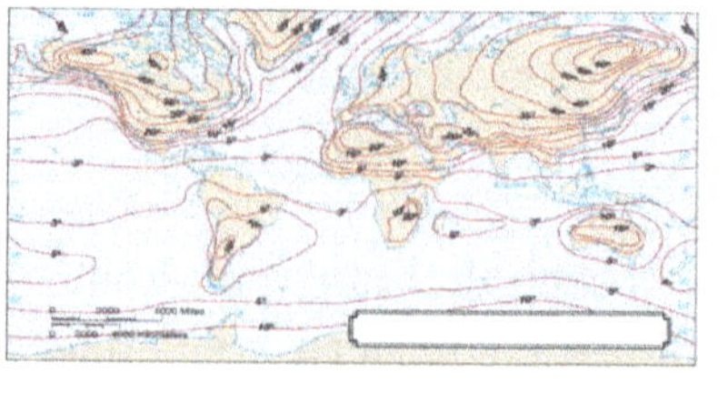

Mean Annual Temperature Range (°C).

Equatorial Trough and Inter Tropical Convergence Zone (ITCZ)

On or near the equator, where average solar radiation is greatest, air is warmed at the surface and rises. This creates a band of low air pressure, centered on the equator. This rising air comprises one segment of a circulation pattern called the Hadley Cell. The rising air is replaced by the Trade winds approaching the equator from north and south. As the trade winds meet near the equator, surface convergence and uplift take place. Equatorial Trough is the quasi-continuous belt of low pressure lying between the subtropical high pressure belts of the Northern and Southern Hemispheres. This entire region is one of very homogeneous air, probably the most ideally barotropic region of the atmosphere. Yet humidity is so high that slight variations in stability cause major variations in weather. The position of the equatorial trough is fairly constant in the eastern portions of the Atlantic and Pacific, but it varies greatly with season in the western portions of those oceans and in southern Asia and the Indian Ocean. It moves into or toward the summer hemisphere. The equatorial trough would be said to contain regions of doldrums; portions of it could be described as intertropical convergence zone (ITCZ).

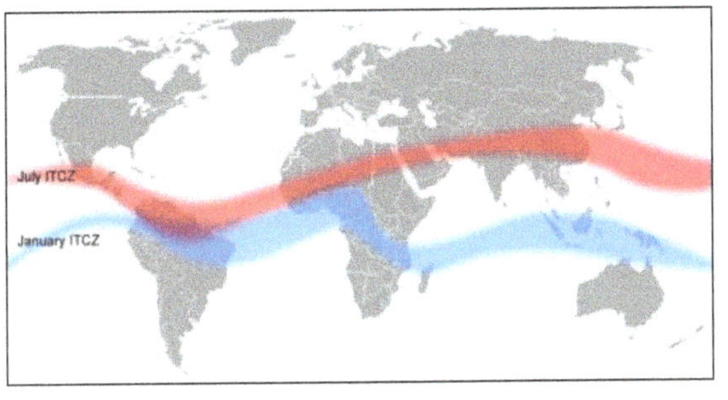

The convergence of the Southeast and Northeast Trade Winds, within the doldrums, creates a zone of Cumulus clouds and attendant shower activity. Under certain circumstances, tropical depressions on the ITCZ intensify to hurricanes. It may seem puzzling that the ITCZ can produce cyclones, when the Coriolis force is at its weakest near the equator. The answer to this puzzle lies in the fact that the ITCZ is not stationary on the equator, but migrates north and south with the seasons. These movements are not perfectly symmetrical above and below the equator, because of the influence of land masses, among other factors.

Climatic Zones

The Köppen Climate Classification System is the most widely used for classifying the world's climates. Köppen's scheme to classify world climates was devised in 1918 by Wladimir Köppen of the University of Graz in Austria. It is based on the concept that native vegetation is the best expression of climate, climate zone boundaries having been selected with vegetation limits in mind. Among the various types of classification of climate, Köppen's classification system is the most widely used for teaching and studying of Geography. A unique and distinctive feature of the system is the employment of an ingenious symbolic nomenclature in designating the climatic types. The classification may be applied to present-day climatic conditions. Alternatively, it also may be used to develop a future climatology that is implied by the output of a numerical climate model.

Köppen recognizes five principal groups of world climates that are intended to correspond with five principal vegetation groups. These five climatic groups may be described as tropical rainy, dry, warm temperate rainy, cold snow-forest, and polar. Each type of climate is described by a formula consisting of a combination of letters, and each of which has a precise meaning. Each of these climates is further divided into sub-divisions based upon differences in the seasonal distribution of temperature and precipitation. This is the most widely used systems for classifying climate because it is easy to use and data requirements are minimal. The classification is based on the 26 values of monthly and annual average precipitation and temperature:

Precipitation (mm) (annual mean) : Jan to Dec (12 monthly values), r

Temperature (^{o}C) (annual mean) : Jan to Dec (12 monthly values), t

The Köppen classification includes five major categories which are designated by capital letters as :

A : Tropical rainy Climates

B : Dry Climates

C : Warm temperate rainy Climates

D : Cold snow-forest Climates

E : Polar Climates

References

- Sene, kevin (2015). Hydrometeorology: forecasting and applications. Springer international publishing switzerland. P. 1. Isbn 978-3-319-23546-2
- Meteorology, encyclopedia: nationalgeographic.org, Retrieved 2 June, 2019
- Dale, murray; davies, paul; harrison, tim (2012). "review of recent advances in uk operational hydrometeorology". Proceedings of the institution of civil engineers : water management. 165 (2): 55–64. Doi:10.1680/wama.2012.165.2.55
- The-different-branches-of-atmospheric-science: worldatlas.com, Retrieved 29 March, 2019
- "flood forecasting centre". Ffc-environment-agency.metoffice.gov.uk. 2011-09-22. Retrieved 2014-05-28
- Climatology, course3, lecture-notes: imd.gov.in, Retrieved 22 May, 2019
- "Republic hydrometeorological service of serbia kneza višeslava 66 beograd". Hidmet.gov.rs. 2014-05-18. Retrieved 2014-05-28

Oceanography

4

- **Geological Oceanography**
- **Physical Oceanography**
- **Chemical Oceanography**

Oceanography is the Earth science that deals with the study of the physical and biological aspects of oceans. The branches of oceanography include geological oceanography, chemical oceanography and physical oceanography. This chapter provides comprehensive insights into oceanography and these branches.

Oceanography is the scientific discipline concerned with all aspects of the world's oceans and seas, including their physical and chemical properties, their origin and geologic framework, and the life forms that inhabit the marine environment.

Traditionally, oceanography has been divided into four separate but related branches: physical oceanography, chemical oceanography, marine geology, and marine ecology. Physical oceanography deals with the properties of seawater (temperature, density, pressure, and so on), its movement (waves, currents, and tides), and the interactions between the ocean waters and the atmosphere.

Chemical oceanography has to do with the composition of seawater and the biogeochemical cycles that affect it. Marine geology focuses on the structure, features, and evolution of the ocean basins. Marine ecology, also called biological oceanography, involves the study of the plants and animals of the sea, including life cycles and food production.

Oceanography is the sum of these several branches. Oceanographic research entails the sampling of seawater and marine life for close study, the remote sensing of oceanic processes with aircraft and Earth-orbiting satellites, and the exploration of the seafloor by means of deep-sea drilling and seismic profiling of the terrestrial crust below the ocean bottom.

Greater knowledge of the world's oceans enables scientists to more accurately predict, for example, long-term weather and climatic changes and also leads to more efficient exploitation of the Earth's resources. Oceanography also is vital to understanding the effect of pollutants on ocean waters and to the preservation of the quality of the oceans' waters in the face of increasing human demands made on them.

Geological Oceanography

Geological oceanography is the study of Earth beneath the oceans. A geological oceanographer studies the topography, structure, and geological processes of the ocean floor to discover how the Earth and oceans were formed and how ongoing processes may change them in the future. Geological oceanography is one of the broadest fields in the Earth Sciences and contains many subdisciplines, including geophysics and plate tectonics, petrology and sedimentation processes, and micropaleontology and stratigraphy. Geological oceanographers study many features of the oceans such as rises and ridges, trenches, seamounts, abyssal hills, the oceanic crust, sedimentation (clastic, chemical, and biological), erosional processes, volcanism, and seismicity.

Tools and Techniques

Many different tools are used by geological oceanographers. For example, the structure and topography of the ocean floor are studied through the use of satellite mapping, which measures the level of the ocean surface to estimate the shape of the ocean floor. A detailed study in 1978 by the National Oceanic and Atmospheric Administration (NOAA) involved the Seasat satellite which produced frequent, short pulses of microwave radiation to measure the level of the surface of the sea with great accuracy (within 3 centimeters or 1 inch). Underwater mountains and valleys cause subtle variations in Earth's gravitational field. The stronger gravity near high massive formations attracts more water molecules, raising the level of the ocean slightly. Similarly valleys on the ocean floor produce weaker areas of gravity, so the level of the ocean will be lower. Using this technique, a complete survey of the ocean floor was accomplished.

Seismic techniques are used to measure the subsurface structure. This type of study is carried out by teams of two ships: one fires an explosive in the water and the other uses sensitive instruments to record the sound waves as they reach the second ship. Some waves travel directly to the second ship; others travel to the ocean floor, are refracted (bent) within the layers of sediment, and then travel to the second ship. By measuring the time it takes for the energy to arrive and the distance between the boats, the thickness of sediments and other features can be determined. Structures may also be analyzed by studying natural earthquake waves that travel through deeper oceanic rocks and may be recorded at stations around the world.

Geological oceanography and coastal geology are closely related yet distinct disciplines. Geological oceanographers study the rocks and structure of the ocean floor, the oceanfloor sediment that covers them, and the processes that formed them. Coastal geologists focus on these structures and processes in a coastal environment.

Other ways oceanic sediments are studied are by dredging processes and deep-sea exploration projects such as the Ocean Drilling Program (ODP), which obtains samples of seafloor sediment from the entire world. The ocean sediments are found to consist of rock particles and organic remains whose compositions depend on depth, distance from continents, and local variations such as submarine volcanoes or high biological activity. For example, clay minerals, which are formed by the weathering of continental rocks, are carried out to sea by rivers and are usually abundant in the deep sea. Thick deposits of weathered rock material, commonly coarser than deep-sea clays, are often found near the mouths of rivers and on the continental shelves.

Physical Oceanography

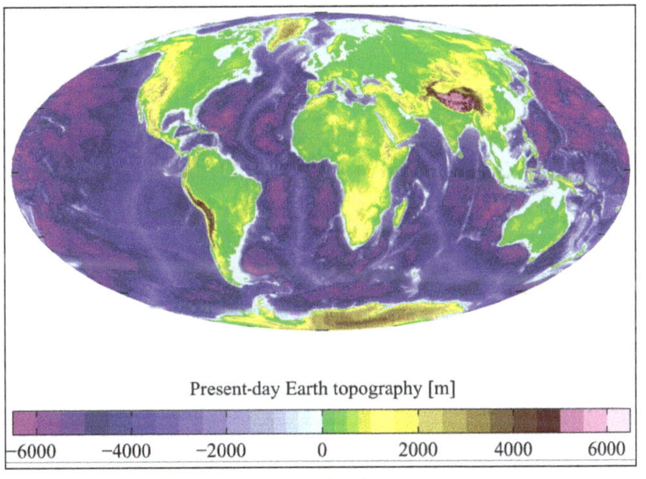

World ocean bathymetry.

Physical oceanography is the study of physical conditions and physical processes within the ocean, especially the motions and physical properties of ocean waters.

It is one of several sub-domains into which oceanography is divided. Others include biological, chemical and geological oceanography. It may be subdivided into descriptive and dynamical physical oceanography.

Descriptive physical oceanography seeks to research the ocean through observations and complex numerical models, which describe the fluid motions as precisely as possible.

Dynamical physical oceanography focuses primarily upon the processes that govern the motion of fluids with emphasis upon theoretical research and numerical models. These are part of the large field of Geophysical Fluid Dynamics (GFD) that is shared together with meteorology. GFD is a sub field of Fluid dynamics describing flows occurring on spatial and temporal scales that are greatly influenced by the Coriolis force.

Physical Setting

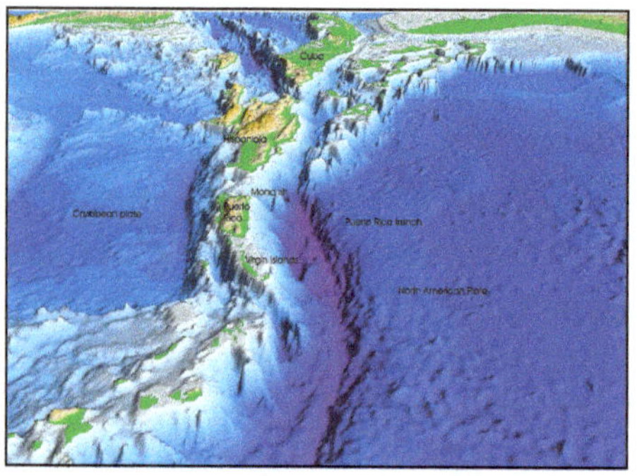

Perspective view of the sea floor of the Atlantic Ocean and the Caribbean Sea. The purple sea floor at the center of the view is the Puerto Rico Trench.

Roughly 97% of the planet's water is in its oceans, and the oceans are the source of the vast majority of water vapor that condenses in the atmosphere and falls as rain or snow on the continents. The tremendous heat capacity of the oceans moderates the planet's climate, and its absorption of various gases affects the composition of the atmosphere. The ocean's influence extends even to the composition of volcanic rocks through seafloor metamorphism, as well as to that of volcanic gases and magmas created at subduction zones.

The oceans are far deeper than the continents are tall; examination of the Earth's hypsographic curve shows that the average elevation of Earth's landmasses is only 840 metres (2,760 ft), while the ocean's average depth is 3,800 metres (12,500 ft). Though this apparent discrepancy is great, for both land and sea, the respective extremes such as mountains and trenches are rare.

Area, volume plus mean and maximum depths of oceans (excluding adjacent seas)				
Body	Area ($10^6 km^2$)	Volume ($10^6 km^3$)	Mean depth (m)	Maximum (m)
Pacific Ocean	165.2	707.6	4282	-11033
Atlantic Ocean	82.4	323.6	3926	-8605
Indian Ocean	73.4	291.0	3963	-8047
Southern Ocean	20.3			-7235
Arctic Ocean	14.1		1038	
Caribbean Sea	2.8			-7686

Temperature, Salinity and Density

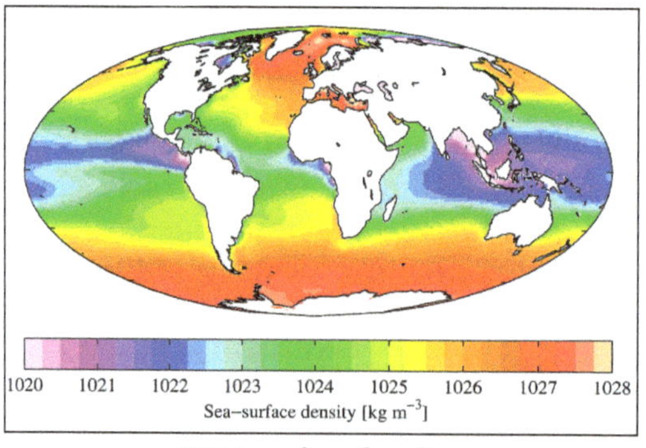

WOA surface density.

Because the vast majority of the world ocean's volume is deep water, the mean temperature of seawater is low; roughly 75% of the ocean's volume has a temperature from 0° – 5 °C. The same percentage falls in a salinity range between 34–35 ppt (3.4–3.5%) (Pinet 1996). There is still quite a bit of variation, however. Surface temperatures can range from below freezing near the poles to 35 °C in restricted tropical seas, while salinity can vary from 10 to 41 ppt (1.0–4.1%).

The vertical structure of the temperature can be divided into three basic layers, a surface mixed layer, where gradients are low, a thermocline where gradients are high, and a poorly stratified abyss.

In terms of temperature, the ocean's layers are highly latitude-dependent; the thermocline is pronounced in the tropics, but nonexistent in polar waters. The halocline usually lies near the surface, where evaporation raises salinity in the tropics, or meltwater dilutes it in polar regions. These variations of salinity and temperature with depth change the density of the seawater, creating the pycnocline.

Circulation

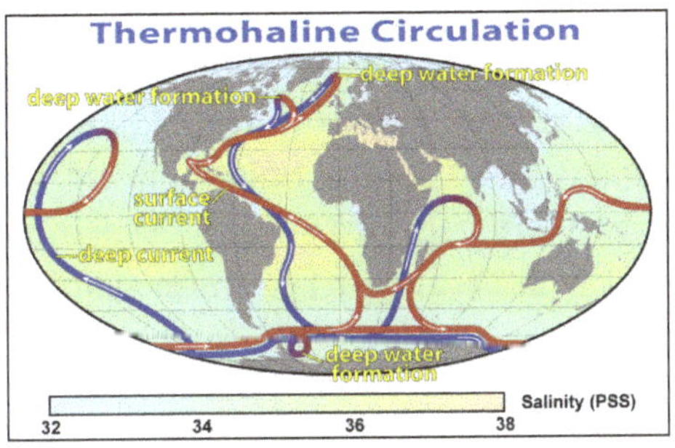

Density-driven thermohaline circulation.

Energy for the ocean circulation (and for the atmospheric circulation) comes from solar radiation and gravitational energy from the sun and moon. The amount of sunlight absorbed at the surface varies strongly with latitude, being greater at the equator than at the poles, and this engenders fluid motion in both the atmosphere and ocean that acts to redistribute heat from the equator towards the poles, thereby reducing the temperature gradients that would exist in the absence of fluid motion. Perhaps three quarters of this heat is carried in the atmosphere; the rest is carried in the ocean.

The atmosphere is heated from below, which leads to convection, the largest expression of which is the Hadley circulation. By contrast the ocean is heated from above, which tends to suppress convection. Instead ocean deep water is formed in polar regions where cold salty waters sink in fairly restricted areas. This is the beginning of the thermohaline circulation.

Oceanic currents are largely driven by the surface wind stress; hence the large-scale atmospheric circulation is important to understanding the ocean circulation. The Hadley circulation leads to Easterly winds in the tropics and Westerlies in mid-latitudes. This leads to slow equatorward flow throughout most of a subtropical ocean basin (the Sverdrup balance). The return flow occurs in an intense, narrow, poleward western boundary current. Like the atmosphere, the ocean is far wider than it is deep, and hence horizontal motion is in general much faster than vertical motion. In the southern hemisphere there is a continuous belt of ocean, and hence the mid-latitude westerlies force the strong Antarctic Circumpolar Current. In the northern hemisphere the land masses prevent this and the ocean circulation is broken into smaller gyres in the Atlantic and Pacific basins.

Coriolis Effect

The Coriolis effect results in a deflection of fluid flows (to the right in the Northern Hemisphere and left in the Southern Hemisphere). This has profound effects on the flow of the oceans. In particular it means the flow goes around high and low pressure systems, permitting them to persist for long periods of time. As a result, tiny variations in pressure can produce measurable currents. A slope of one part in one million in sea surface height, for example, will result in a current of 10 cm/s at mid-latitudes. The fact that the Coriolis effect is largest at the poles and weak at the equator results in sharp, relatively steady western boundary currents which are absent on eastern boundaries.

Ekman Transport

Ekman transport results in the net transport of surface water 90 degrees to the right of the wind in the Northern Hemisphere, and 90 degrees to the left of the wind in the Southern Hemisphere. As the wind blows across the surface of the ocean, it "grabs" onto a thin layer of the surface water. In turn, that thin sheet of water transfers motion energy to the thin layer of water under it, and so on. However, because of the Coriolis Effect, the direction of travel of the layers of water slowly move farther and farther to the right as they get deeper in the Northern Hemisphere, and to the left in the Southern Hemisphere. In most cases, the very bottom layer of water affected by the wind is at a depth of 100 m – 150 m and is traveling about 180 degrees, completely opposite of the direction that the wind is blowing. Overall, the net transport of water would be 90 degrees from the original direction of the wind.

Langmuir Circulation

Langmuir circulation results in the occurrence of thin, visible stripes, called windrows on the surface of the ocean parallel to the direction that the wind is blowing. If the wind is blowing with more than 3 m s^{-1}, it can create parallel windrows alternating upwelling and downwelling about 5–300 m apart. These windrows are created by adjacent ovular water cells (extending to about 6 m (20 ft) deep) alternating rotating clockwise and counterclockwise. In the convergence zones debris, foam and seaweed accumulates, while at the divergence zones plankton are caught and carried to the surface. If there are many plankton in the divergence zone fish are often attracted to feed on them.

Ocean–Atmosphere Interface

Hurricane Isabel east of the Bahamas on 15 September 2003

At the ocean-atmosphere interface, the ocean and atmosphere exchange fluxes of heat, moisture and momentum.

Heat

The important heat terms at the surface are the sensible heat flux, the latent heat flux, the incoming solar radiation and the balance of long-wave (infrared) radiation. In general, the tropical oceans will tend to show a net gain of heat, and the polar oceans a net loss, the result of a net transfer of energy polewards in the oceans.

The oceans' large heat capacity moderates the climate of areas adjacent to the oceans, leading to a maritime climate at such locations. This can be a result of heat storage in summer and release in winter; or of transport of heat from warmer locations: a particularly notable example of this is Western Europe, which is heated at least in part by the north atlantic drift.

Momentum

Surface winds tend to be of order meters per second; ocean currents of order centimeters per second. Hence from the point of view of the atmosphere, the ocean can be considered effectively stationary; from the point of view of the ocean, the atmosphere imposes a significant wind stress on its surface, and this forces large-scale currents in the ocean.

Through the wind stress, the wind generates ocean surface waves; the longer waves have a phase velocity tending towards the wind speed. Momentum of the surface winds is transferred into the energy flux by the ocean surface waves. The increased roughness of the ocean surface, by the presence of the waves, changes the wind near the surface.

Moisture

The ocean can gain moisture from rainfall, or lose it through evaporation. Evaporative loss leaves the ocean saltier; the Mediterranean and Persian Gulf for example have strong evaporative loss; the resulting plume of dense salty water may be traced through the Straits of Gibraltar into the Atlantic Ocean. At one time, it was believed that evaporation/precipitation was a major driver of ocean currents; it is now known to be only a very minor factor.

Planetary Waves

Kelvin Waves

A Kelvin wave is any progressive wave that is channeled between two boundaries or opposing forces (usually between the Coriolis force and a coastline or the equator). There are two types, coastal and equatorial. Kelvin waves are gravity driven and non-dispersive. This means that Kelvin waves can retain their shape and direction over long periods of time. They are usually created by a sudden shift in the wind, such as the change of the trade winds at the beginning of the El Niño-Southern Oscillation.

Coastal Kelvin waves follow shorelines and will always propagate in a counterclockwise direction in the Northern hemisphere (with the shoreline to the right of the direction of travel) and clockwise in the Southern hemisphere.

Equatorial Kelvin waves propagate to the east in the Northern and Southern hemispheres, using the equator as a guide.

Kelvin waves are known to have very high speeds, typically around 2–3 meters per second. They have wavelengths of thousands of kilometers and amplitudes in the tens of meters.

Rossby Waves

Rossby waves, or planetary waves are huge, slow waves generated in the troposphere by temperature differences between the ocean and the continents. Their major restoring force is the change in Coriolis force with latitude. Their wave amplitudes are usually in the tens of meters and very large wavelengths. They are usually found at low or mid latitudes.

There are two types of Rossby waves, barotropic and baroclinic. Barotropic Rossby waves have the highest speeds and do not vary vertically. Baroclinic Rossby waves are much slower.

The special identifying feature of Rossby waves is that the phase velocity of each individual wave always has a westward component, but the group velocity can be in any direction. Usually the shorter Rossby waves have an eastward group velocity and the longer ones have a westward group velocity.

Climate Variability

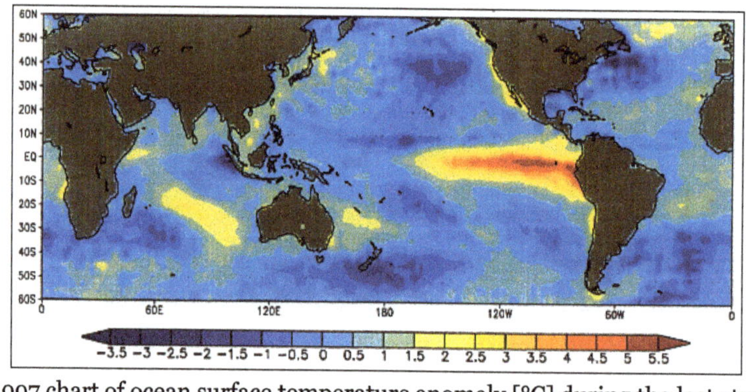

December 1997 chart of ocean surface temperature anomaly [°C] during the last strong El Niño

The interaction of ocean circulation, which serves as a type of heat pump, and biological effects such as the concentration of carbon dioxide can result in global climate changes on a time scale of decades. Known climate oscillations resulting from these interactions, include the Pacific decadal oscillation, North Atlantic oscillation, and Arctic oscillation. The oceanic process of thermohaline circulation is a significant component of heat redistribution across the globe, and changes in this circulation can have major impacts upon the climate.

Antarctic Circumpolar Wave

This is a coupled ocean/atmosphere wave that circles the Southern Ocean about every eight years. Since it is a wave-2 phenomenon (there are two peaks and two troughs in a latitude circle) at each fixed point in space a signal with a period of four years is seen. The wave moves eastward in the direction of the Antarctic Circumpolar Current.

Ocean Currents

Among the most important ocean currents are the:

- Antarctic Circumpolar Current
- Deep ocean (density-driven)
- Western boundary currents
    - Gulf Stream
    - Kuroshio Current
    - Labrador Current
    - Oyashio Current
    - Agulhas Current
    - Brazil Current
    - East Australia Current
- Eastern Boundary currents
    - California Current
    - Canary Current
    - Peru Current
    - Benguela Current

Antarctic Circumpolar

The ocean body surrounding the Antarctic is currently the only continuous body of water where there is a wide latitude band of open water. It interconnects the Atlantic, Pacific and Indian oceans, and provide an uninterrupted stretch for the prevailing westerly winds to significantly increase wave amplitudes. It is generally accepted that these prevailing winds are primarily responsible for the circumpolar current transport. This current is now thought to vary with time, possibly in an oscillatory manner.

Deep Ocean

In the Norwegian Sea evaporative cooling is predominant, and the sinking water mass, the North Atlantic Deep Water (NADW), fills the basin and spills southwards through crevasses in the submarine sills that connect Greenland, Iceland and Britain. It then flows along the western boundary of the Atlantic with some part of the flow moving eastward along the equator and then poleward into the ocean basins. The NADW is entrained into the Circumpolar Current, and can be traced into the Indian and Pacific basins. Flow from the Arctic Ocean Basin into the Pacific, however, is blocked by the narrow shallows of the Bering Strait.

Western Boundary

An idealised subtropical ocean basin forced by winds circling around a high pressure (anticyclonic) systems such as the Azores-Bermuda high develops a gyre circulation with slow steady flows towards the equator in the interior. As discussed by Henry Stommel, these flows are balanced in the region of the western boundary, where a thin fast polewards flow called a western boundary current develops. Flow in the real ocean is more complex, but the Gulf stream, Agulhas and Kuroshio are examples of such currents. They are narrow (approximately 100 km across) and fast (approximately 1.5 m/s).

Equatorwards western boundary currents occur in tropical and polar locations, e.g. the East Greenland and Labrador currents, in the Atlantic and the Oyashio. They are forced by winds circulation around low pressure (cyclonic).

Gulf Stream

The Gulf Stream, together with its northern extension, North Atlantic Current, is a powerful, warm, and swift Atlantic Ocean current that originates in the Gulf of Mexico, exits through the Strait of Florida, and follows the eastern coastlines of the United States and Newfoundland to the northeast before crossing the Atlantic Ocean.

Kuroshio

The Kuroshio Current is an ocean current found in the western Pacific Ocean off the east coast of Taiwan and flowing northeastward past Japan, where it merges with the easterly drift of the North Pacific Current. It is analogous to the Gulf Stream in the Atlantic Ocean, transporting warm, tropical water northward towards the polar region.

Heat flux

Heat Storage

Ocean heat flux is a turbulent and complex system which utilizes atmospheric measurement techniques such as eddy covariance to measure the rate of heat transfer expressed in the unit of joules or watts per second. Heat flux is the difference in temperature between two points through which the heat passes. Most of the Earth's heat storage is within its seas with smaller fractions of the heat transfer in processes such as evaporation, radiation, diffusion, or absorption into the sea floor. The majority of the ocean heat flux is through advection or the movement of the ocean's currents. For example, the majority of the warm water movement in the south Atlantic is thought to have originated in the Indian Ocean. Another example of advection is the nonequatorial Pacific heating which results from subsurface processes related to atmospheric anticlines. Recent warming observations of Antarctic Bottom Water in the Southern Ocean is of concern to ocean scientists because bottom water changes will effect currents, nutrients, and biota elsewhere. The international awareness of global warming has focused scientific research on this topic since the 1988 creation of the Intergovernmental Panel on Climate Change. Improved ocean observation, instrumentation, theory, and funding has increased scientific reporting on regional and global issues related to heat.

Sea Level Change

Tide gauges and satellite altimetry suggest an increase in sea level of 1.5–3 mm/yr over the past 100 years.

The IPCC predicts that by 2081-2100, global warming will lead to a sea level rise of 260 to 820 mm.

Rapid Variations

Tides

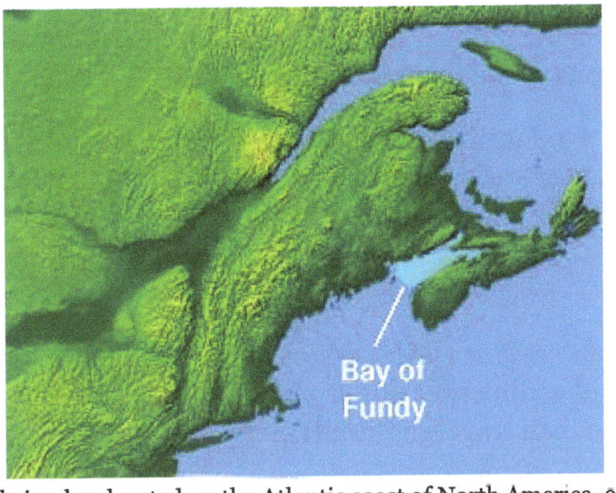

The Bay of Fundy is a bay located on the Atlantic coast of North America, on the northeast end of the Gulf of Maine between the provinces of New Brunswick and Nova Scotia.

The rise and fall of the oceans due to tidal effects is a key influence upon the coastal areas. Ocean tides on the planet Earth are created by the gravitational effects of the Sun and Moon. The tides produced by these two bodies are roughly comparable in magnitude, but the orbital motion of the Moon results in tidal patterns that vary over the course of a month.

The ebb and flow of the tides produce a cyclical current along the coast, and the strength of this current can be quite dramatic along narrow estuaries. Incoming tides can also produce a tidal bore along a river or narrow bay as the water flow against the current results in a wave on the surface.

Tide and Current clearly illustrates the impact of these natural cycles on the lifestyle and livelihood of Native Hawaiians tending coastal fishponds.

Tidal resonance occurs in the Bay of Fundy since the time it takes for a large wave to travel from the mouth of the bay to the opposite end, then reflect and travel back to the mouth of the bay coincides with the tidal rhythm producing the world's highest tides.

As the surface tide oscillates over topography, such as submerged seamounts or ridges, it generates internal waves at the tidal frequency, which are known as internal tides.

Tsunamis

A series of surface waves can be generated due to large-scale displacement of the ocean water.

These can be caused by sub-marine landslides, seafloor deformations due to earthquakes, or the impact of a large meteorite.

The waves can travel with a velocity of up to several hundred km/hour across the ocean surface, but in mid-ocean they are barely detectable with wavelengths spanning hundreds of kilometers.

Tsunamis, originally called tidal waves, were renamed because they are not related to the tides. They are regarded as shallow-water waves, or waves in water with a depth less than 1/20 their wavelength. Tsunamis have very large periods, high speeds, and great wave heights.

The primary impact of these waves is along the coastal shoreline, as large amounts of ocean water are cyclically propelled inland and then drawn out to sea. This can result in significant modifications to the coastline regions where the waves strike with sufficient energy.

The tsunami that occurred in Lituya Bay, Alaska on July 9, 1958 was 520 m (1,710 ft) high and is the biggest tsunami ever measured, almost 90 m (300 ft) taller than the Sears Tower in Chicago and about 110 m (360 ft) taller than the former World Trade Center in New York.

Surface Waves

The wind generates ocean surface waves, which have a large impact on offshore structures, ships, coastal erosion and sedimentation, as well as harbours. After their generation by the wind, ocean surface waves can travel (as swell) over long distances.

Chemical Oceanography

Chemical oceanography investigates the chemical composition of sea water. Chemical oceanographers study the interactions between organic and inorganic substances and the biological, physical and geological conditions of the ocean. They want to understand how marine chemistry is influenced by physical processes and exchanges with the atmosphere, biosphere and geosphere.

Processes, which introduce chemical species into the ocean and those which remove or transform the substances are important targets of marine chemical research. One preferentially investigate the cycling of those elements which are important for biological processes (eg. carbon, nitrogen, phosphorus, iron). The ocean contains each chemical element of our planet either in its pure form or in any kind of chemical compounds either dissolved or particulate.

Besides the main components of seawater, which strongly dictate its properties, many elements and compounds occur as trace ingredients confering great biogeochemical interest. These constituents often exhibit important biological or chemical functions, which stimulate the specific interests of marine chemistry and chemical oceanography. Sea water contains many gaseous components, which are introduced into the atmosphere or are absorbed from it.

The origin and fate of gases, affecting the atmosphere are a very special element of research therefore. Green house gases, contributing to atmospheric warming, trace gases forming aerosols and those destroying ozone belong to this category.

References

- Oceanography, science: britannica.com, Retrieved 10 July, 2019
- D., talley, lynne; l., pickard, george; j., emery, william; (oceanographer), swift, james h. Descriptive physical oceanography : an introduction. Isbn 9780750645522. Oclc 784140610
- Oceanography-geological, oc-po: waterencyclopedia.com, Retrieved 29 June, 2019
- Marshak, stephen (2001). Earth: portrait of a planet. New york: w.w. norton & company. Isbn 0-393-97423-5
- Fb2-ch, fb2, research: geomar.de, Retrieved 30 January, 2019

Hydrology 5

• Hydrometry

Hydrology is a discipline that studies the distribution, quality and movement of water on Earth and other planets. It focuses on the study of water resources, water cycle and environmental watershed sustainability. The diverse applications of hydrology as well as its sub-domains such as hydrometeorology and hydrometry have been thoroughly discussed in this chapter.

Water is one of our most important natural resources. Without it, there would be no life on earth. The supply of water available for our use is limited by nature. Although there is plenty of water on earth, it is not always in the right place, at the right time and of the right quality. Adding to the problem is the increasing evidence that chemical wastes improperly discarded yesterday are showing up in our water supplies today. Hydrology has evolved as a science in response to the need to understand the complex water systems of the Earth and help solve water problems. Hydrologists play a vital role in finding solutions to water problems, and interesting and challenging careers are available to those who choose to study hydrology.

Water and People

Estimates of water use in the United States indicate that about 355 billion gallons per day (one thousand million gallons per day, abbreviated Bgal/d) were withdrawn for all uses during 2010. This total has declined about 17 percent since 1980. Fresh groundwater withdrawals (76.0 Bgal/d) during 2010 were 8 percent less than during 1980. Fresh surface-water withdrawals for 2010 were 230 Bgal/d, 18 percent less than in 1980.

Much of our water use is hidden. Think about what you had for lunch. A hamburger, for example, requires water to raise wheat for the bun, to grow hay and corn to feed the cattle and to process the bread and beef. Together with french fries and a soft drink, this all-American meal uses about 1,500 gallons of water — enough to fill a small swimming pool. How about your clothes? To grow cotton for a pair of jeans takes about 400 gallons. A shirt requires about 400 gallons. How do you get to school or to the store? To produce the amount of finished steel in a car has in the past required about 32,000 gallons of water. Similarly, the steel in a 30-pound bicycle required 480 gallons. This shows that industry must continue to strive to reduce water use through manufacturing processes that use less water, and through recycling of water.

Hydrology is the science that encompasses the occurrence, distribution, movement and properties of the waters of the earth and their relationship with the environment within each phase of the hydrologic cycle. The water cycle, or hydrologic cycle, is a continuous process by which water is purified by evaporation and transported from the earth's surface (including the oceans) to the atmosphere and back to the land and oceans. All of the physical, chemical and biological processes involving water as it travels its various paths in the atmosphere, over and beneath the earth's surface and through growing plants, are of interest to those who study the hydrologic cycle.

There are many pathways the water may take in its continuous cycle of falling as rainfall or snowfall and returning to the atmosphere. It may be captured for millions of years in polar ice caps. It may flow to rivers and finally to the sea. It may soak into the soil to be evaporated directly from the soil surface as it dries or be transpired by growing plants. It may percolate through the soil to ground water reservoirs (aquifers) to be stored or it may flow to wells or springs or back to streams by seepage. The cycle for water may be short, or it may take millions of years.

People tap the water cycle for their own uses. Water is diverted temporarily from one part of the cycle by pumping it from the ground or drawing it from a river or lake. It is used for a variety of activities such as households, businesses and industries; for irrigation of farms and parklands; and for production of electric power. After use, water is returned to another part of the cycle: perhaps discharged downstream or allowed to soak into the ground. Used water normally is lower in quality, even after treatment, which often poses a problem for downstream users.

The hydrologist studies the fundamental transport processes to be able to describe the quantity and quality of water as it moves through the cycle (evaporation, precipitation, streamflow, infiltration, ground water flow, and other components). The engineering hydrologist, or water resources engineer, is involved in the planning, analysis, design, construction and operation of projects for the control, utilization, and management of water resources. Water resources problems are also the concern of meteorologists, oceanographers, geologists, chemists, physicists, biologists, economists, political scientists, specialists in applied mathematics and computer science, and engineers in several fields.

Surface Water

Most cities meet their needs for water by withdrawing it from the nearest river, lake or reservoir. Hydrologists help cities by collecting and analyzing the data needed to predict how much water is available from local supplies and whether it will be sufficient to meet the city's projected future needs. To do this, hydrologists study records of rainfall, snowpack depths and river flows that are collected and compiled by hydrologists in various government agencies. They inventory the extent river flow already is being used by others.

Managing reservoirs can be quite complex, because they generally serve many purposes. Reservoirs increase the reliability of local water supplies. Hydrologists use topographic maps and aerial photographs to determine where the reservoir shorelines will be and to calculate reservoir depths and storage capacity. This work ensures that, even at maximum capacity, no highways, railroads or homes would be flooded.

Deciding how much water to release and how much to store depends upon the time of year, flow predictions for the next several months, and the needs of irrigators and cities as well as downstream water-users that rely on the reservoir. If the reservoir also is used for recreation or for generation of hydroelectric power, those requirements must be considered. Decisions must be coordinated with other reservoir managers along the river. Hydrologists collect the necessary information, enter it into a computer, and run computer models to predict the results under various operating strategies. On the basis of these studies, reservoir managers can make the best decision for those involved.

The availability of surface water for swimming, drinking, industrial or other uses sometimes is restricted because of pollution. Pollution can be merely an unsightly and inconvenient nuisance, or it can be an invisible, but deadly, threat to the health of people, plants and animals.

Hydrologists assist public health officials in monitoring public water supplies to ensure that health standards are met. When pollution is discovered, environmental engineers work with hydrologists in devising the necessary sampling program. Water quality in estuaries, streams, rivers and lakes must be monitored, and the health of fish, plants and wildlife along their stretches surveyed. Related work concerns acid rain and its effects on aquatic life, and the behavior of toxic metals and organic chemicals in aquatic environments. Hydrologic and water quality mathematical models are developed and used by hydrologists for planning and management and predicting water quality effects of changed conditions. Simple analyses such as pH, turbidity, and oxygen content may be done by hydrologists in the field. Other chemical analyses require more sophisticated laboratory equipment. In the past, municipal and industrial sewage was a major source of pollution for streams and lakes. Such wastes often received only minimal treatment, or raw wastes were dumped into rivers. Today, we are more aware of the consequences of such actions, and billions of dollars must be invested in pollution-control equipment to protect the waters of the earth. Other sources of pollution are more difficult to identify and control. These include road deicing salts, storm runoff from urban areas and farmland, and erosion from construction sites.

Groundwater

Groundwater, pumped from beneath the earth's surface, is often cheaper, more convenient and less vulnerable to pollution than surface water. Therefore, it is commonly used for public water supplies. Groundwater provides the largest source of usable water storage in the United States. Underground reservoirs contain far more water than the capacity of all surface reservoirs and lakes, including the Great Lakes. In some areas, ground water may be the only option. Some municipalities survive solely on groundwater.

Hydrologists estimate the volume of water stored underground by measuring water levels in local wells and by examining geologic records from well-drilling to determine the extent, depth and thickness of water-bearing sediments and rocks. Before an investment is made in full-sized wells, hydrologists may supervise the drilling of test wells. They note the depths at which water is encountered and collect samples of soils, rock and water for laboratory analyses. They may run a variety of geophysical tests on the completed hole, keeping and accurate log of their observations and test results. Hydrologists determine the most efficient pumping rate by monitoring the extent that water levels drop in the pumped well and in its nearest neighbors. Pumping the well too fast could cause it to go dry or could interfere with neighboring wells. Along the coast, overpumping can cause saltwater intrusion. By plotting and analyzing these data, hydrologists can estimate the maximum and optimum yields of the well.

Polluted ground water is less visible, but more insidious and difficult to clean up, than pollution in rivers and lakes. Ground water pollution most often results from improper disposal of wastes on land. Major sources include industrial and household chemicals and garbage landfills, industrial waste lagoons, tailings and process wastewater from mines, oil field brine pits, leaking underground oil storage tanks and pipelines, sewage sludge and septic systems. Hydrologists provide guidance in the location of monitoring wells around waste disposal sites and sample them at regular intervals to determine if undesirable leachate — contaminated water containing toxic or hazardous chemicals — is reaching the ground water.

In polluted areas, hydrologists may collect soil and water samples to identify the type and extent of contamination. The chemical data then are plotted on a map to show the size and direction of waste movement. In complex situations, computer modeling of water flow and waste migration provides guidance for a clean-up program. In extreme cases, remedial actions may require excavation of the polluted soil. Today, most people and industries realize that the amount of money invested in prevention is far less than that of cleanup. Hydrologists often are consulted for selection of proper sites for new waste disposal facilities. The danger of pollution is minimized by locating wells in areas of deep ground water and impermeable soils. Other practices include lining the bottom of a landfill with watertight materials, collecting any leachate with drains, and keeping the landfill surface covered as much as possible. Careful monitoring is always necessary.

Hydrometry

Hydrometry is the science which is distinct and complementary to hydrology (science of water in its natural environment) and hydraulics (physics of flows), is the discipline that seeks to measure river flows.

The flow rate -volume of water crossing a section of a stream for one unit of time- is expressed in cubic metres per second (m^3/s).

Each watercourse follows a particular regime, determined by the rhythm of precipitation and its hydrological "terroir". For the world's most populated river, the Amazon, the variation in flow between two extreme months of the same month is only one to two. And from one year to the next, its average annual flow at its mouth varies only 10 to 15% around its 206,000 m^3/s value. The Amazon is an extremely regular river.

On the other hand, an African river like the Chari has an average flow of 1197 m^3/s at its outlet in Lake Chad. Within the same year, the variation in flow between two extreme months is a factor of 20 (150 to 3000 m^3/s). And from one year to the next, the average annual flow can vary by a factor of two: 739 m^3/s in 1942, 1720 m^3/s in 1956. The Chari therefore has a much more contrasting regime.

But how are these flows measured? Since ancient times, mankind has been interested in it, at the very least, when it became dependent on agriculture. But it is a much more difficult problem than its familiarity would suggest. What made James Jeans write: "The total radiation emitted by the Sun in the unit of time, transformed into mass, is something like 10,000 times that of the water flowing in the Thames under the London Bridge; and incidentally, if the factor 10,000 is gross, it is not because we do not know the exact mass of the solar radiation, but because we are not able to measure the average flow of the Thames."

Reasons for Measuring River Flow Rates

The measurement of river flows serves several purposes:

- Operational management of hydraulic structures (hydroelectric facilities, irrigation systems, flood control tanks or low water level support systems, etc.);
- the dimensioning of these structures, through knowledge of the characteristics of these watercourses;
- Regulatory control, for verification of flow release obligations downstream of structures (minimum flow to ensure fish survival, maintenance of other uses; non-aggravation of floods), declaration of disaster status (droughts);
- Protection of property and persons, through flood warning;
- Of heritage, by the constitution of series of long-term observations, essential to know the evolution of river regimes, to raise awareness of natural risks, to assign a probability to extreme events (floods, low water levels).

Flow measurement station of the Isère in Grenoble.

The interest of these measures is now reinforced by the current challenges of global warming, new demands for sharing water between different uses (recreational, energy, irrigation, drinking water), the restoration or preservation of natural environments and their biodiversity, the social demand for knowledge, the increased vulnerability of society.

It should be noted that there are currently about 3500 hydrometric stations in metropolitan France, mainly managed by the Ministry of the Environment and by operators of hydroelectric or irrigation works. More than 80% are teletransmitted in real time. This density (0.63/ 100 km^2) is in the average of Western Europe, about the same as that of the United Kingdom, higher than Spain, but lower than Switzerland or Germany.

How is the Flow of Rivers Measured

Direct flow measurement is a complex operation that can only be performed occasionally. Except in very specific cases, direct and continuous monitoring of the flow cannot be carried out. It is the water level that is measured continuously, after having previously connected it to the flow rate by a calibration curve. This is why hydrometry is a 4-step process:

- The continuous measurement of heights upstream of a hydraulic control, or at another location where a unique height-flow relationship can be established,
- The realization of periodic gauges to build this relationship (calibration curve), allowing to convert the heights into flows,
- The layout of this calibration curve and the detection of its evolutions,
- Then, after conversion of heights into flow, critical analysis of spatial and temporal fluctuations, then their archiving.

Hydraulic control in the river. Note the formation of a torrential flow regime as soon as the threshold is crossed, followed by a hydraulic jump producing eddies and a great loss of energy.

Continuously Measure Heights

For a long time, height measurements consisted of visual readings taken daily (or at a shorter frequency) on graduated scales. Over time, the process has become automated by the installation – in addition to these reference scales – of sensors to monitor height variations at a time step adapted to flow fluctuations (very reactive in the case of a small torrential basin; much smoother in a large plain river basin). Several generations of sensors now coexist on the networks: float, pneumatic, piezoresistive, ultra-submerged sound, differential conductivity measurement, etc.

Scale on Niger in Mopti.

All these devices are placed in or in contact with water; the radar – which appeared at the beginning of the 2000s – offers the advantage of being out of the water (a guarantee of better durability, as it is not subject to the aggressions of water, sediments & bodies floating in rivers) and insensitive to temperature (a characteristic that is lacking with emerging ultra sounds). However, the need to move the radar away from the shoreline (edge effects) and the wave reception task conditioned by the waveguide may nevertheless penalize the representativeness of the measurement in relation to the reference scale.

Level measurement by emerged radar.

Calibrate the Calibration Curve: The Gauges

Periodic gauging is carried out on the entire range of flows that the river can reach (in drought, medium water, and floods), mainly by exploring the velocity field or diluting a tracer.

The gauging by exploring the velocity field of the flow has long been limited to surface velocities (by means of “floats“, sticks weighted according to the current). More complete maps of the velocity field are now available using speed sensors: mechanical reels – a propeller rotating in proportion to the local velocity of the current – electromagnetic – the displacement of water producing an induced voltage proportional to the local velocity of the current (Faraday Principle).

Hydrometric reel implemented at Grenoble Campus sur l'Isère station.

Since the early 1990s, ADCP profilers (Acoustic Doppler Current Profiler: a device from oceanography, based on the Doppler effect) have been a real technological leap forward in hydrometry. They significantly reduce on-site measurement time, especially on large rivers, and are now suitable for small rivers (but a minimum depth of 50 cm is required).

Left: Rhodamine injection device. Right: Rhodamine enters the stream. The tracer (harmless to fauna and flora) does not colour the water. Its dilution is then analysed by a fluorimeter (measuring the attenuation of its fluorescence), still today in the laboratory; tomorrow can be directly in the field.

Dilution gauging consists of injecting a tracer in solution into the watercourse and monitoring its concentration over time. When the condition of good tracer mixing is ensured – and if there is no loss of water in the dilution basin – by mass conservation law, the dilution factor is directly proportional to the flow of the river. Several generations of tracers have been historically used, the current state of the art being to favour fluorescent tracers (rhodamine, uranine) or cooking salt.

Linking Height and Flow: The Calibration Curve

The most delicate link is the setting curve, the relationship between height and flow. For a long time manually drawn, according to the operators' expertise alone, the definition of this curve now calls for decision support tools, tools combining statistical approaches, taking into account metrological uncertainties on gauging, hydraulic models.

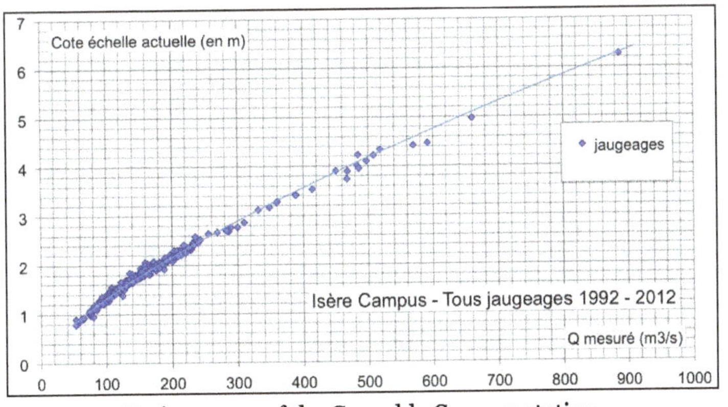

Taring curve of the Grenoble Campus station.

For most river stations, the setting curve can change for multiple reasons and never tends – as a general rule – towards stabilization. Here at Grenoble Campus, the calibration curve shown is the one valid between 1992 & 2012. It was then modified due to work carried out on the Isère dikes from September 2012. The curve evolves rapidly until April 2013 and seems to have stabilized again since.

The height-flow relationship, if it is considered stable over a given period of time, is not necessarily stable over time, especially when the hydraulic control is not constituted by an artificial structure. Vegetation, human intervention, flooding – through the associated mechanisms of solid transport, erosion or deposition – more or less often modify the flow profile of the river. Monitoring the setting curve thus conditions a real gauging strategy, to be adapted both temporally (frequency of gauging) and according to water conditions (low water, average water, floods). Monitoring and plotting the calibration curve is the core business of hydrometry.

The state of the art has recently evolved with indwelling devices that allow continuous speed measurement, either on the surface (speed radar) or indwelling in the flow (transit time ultrasound or Doppler effect). The principles of hydrometry are not fundamentally changed: a calibration relationship of height, velocity(s), flow rate remains to be calibrated throughout the operation of the measurement site. These systems were already implemented when a unique relationship between height and flow was not verified (rivers regulated by navigation and/or subject to tide), but current technological developments make it less costly to distribute this type of installation.

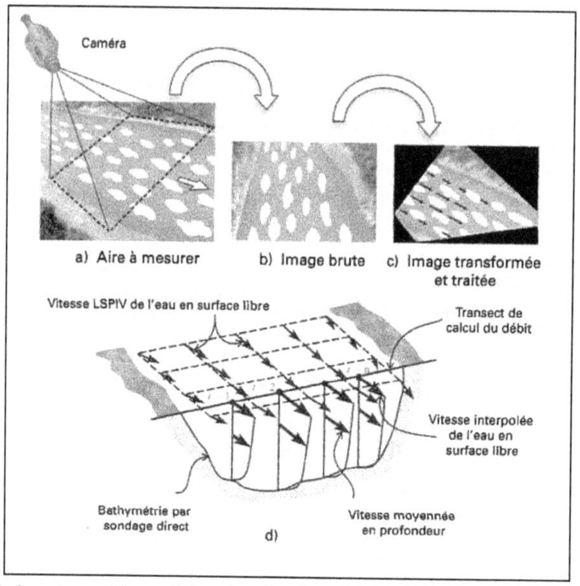

LSPIV principle: a) seeding of floating bodies, b) image recording, c) ortho image rectification, d) flow calculation from LSPIV surface velocity measurements (from Muste).

New imaging technologies bring a promising innovation: video image processing to determine the field of surface velocities of a river. We use here the displacement of all solid bodies transported on the surface (twigs, bubbles, leaves) as well as the turbulence of the flow. This technique is derived from the Particle Image Velocimetry (PIV) used in the laboratory, but for a study on large-scale river-type objects, hence its name Large-Scale PIV (LSPIV). This includes:

- Recording time-stamped image sequences of the flow,
- A Geometric correction of the images to avoid perspective distortions,
- A calculation of the displacement of the flow tracers using a statistical analysis in correlation with the patterns.

Knowing the geometry of the river section and assuming a vertical velocity distribution model, the total flow is estimated from the LSPIV velocity field.

This technique of the future opens the way to a densification of flood measures: the fleeting nature of the episodes, the difficulties of access (flooded roads), the security conditions (violent flows) not allowing the teams to intervene as much as necessary. However, it cannot yet be implemented in case of poor visibility (night, fog).

Check the Consistency of the Data

The conversion of heights into flow, the critique of the results, the archiving in the database are the last part of the hydrometry business.

Consistency tests are carried out on the recordings at the station (identification of shifts & sensor drifts, smoothing of the raw signal, filling in gaps over recording failure periods) and by more or less sophisticated hydrological models:

- In coherence with other upstream and downstream measurement sites.
- With reference to historical data already compiled at the measurement site, by comparing it with previous years, looking for explanations based on the measurement of rainfall, known influences (water withdrawals, etc).

The whole process is iterative, and therefore can lead to questioning the current calibration curve and thus redefine its layout, or even the calibration strategy. Information obtained long after the occurrence of the hydrological event (hydraulic modelling, flood gauging) can lead to significant changes in the results published at a station. It is common to allow a period of eighteen months to two years for the consolidation of information.

The quantification of uncertainties in hydrometry has progressed considerably in recent years, but remains an area of investigation for the profession. It is considered that on the best stations (i.e. those where the calibration curve can be followed at a rate of less than 4 or 5 gauges per year), the current flows – encountered 80% of the time – are consolidated to within 5%.

What are the Current Challenges for Hydrometry

Let us keep in mind that hydrometry is a labour-intensive process, which requires travel in the field and is a real craftsman's task combining metrology, hydraulics and hydrology. As a result, the annual operating cost of a station is often in the order of magnitude of the initial investment cost to create the measurement point. Hydrometry is therefore a long-term task, where budget cuts have a major impact on the quality of the data produced.

Hydrometry is also a complex process, as it affects the natural environment, with all its associated hazards, and where the maturation times of the data can potentially be long. Thus, between information given on the spot (or even used to make a decision) and consolidated data after criticism or discovery of new elements, significant differences may appear (twice as much for an extreme regime value, in flood or drought) several years after their occurrence.

Finally, hydrometry is a process in the making: the availability of new imaging (LSPIV) and communication technologies (telephony, internet) will increase the flow of collected data.

Questions will quickly arise about the processing of this information, its criticism and homogenization, its conservation, and the skills that accompany this massification of information. All this is in response to a real social demand for a better knowledge of the environment, a reduction in vulnerability to hazards, in the current context of climate change and the preservation of biodiversity.

References

- "hydro-meteorological hazards | united nations educational, scientific and cultural organization". Unesco.org. Retrieved 2016-08-05
- What-hydrology?qt-science-center-objects, science, water-science-school, special-topic: usgs.gov, Retrieved 31 March, 2019
- "republic hydrometeorological service of serbia kneza višeslava 66 beograd". Hidmet.gov.rs. 2014-05-18. Retrieved 2014-05-28
- Hydrometry-measuring-flow-river-why-how, water: encyclopedie-environnement.org, Retrieved 14 July, 2019
- "centro nacional de monitoramento e alertas de desastres naturais (cemaden)". Labhidro-igeo-ufrj. Retrieved 2017-06-29

Planetary Science 6

- **Planetary Geology**
- **Cosmochemistry**
- **Space Physics**
- **Solar System**
- **Star System**

Planetary science is the study of the planets and their planetary systems as well as the processes related to their formation. Some of the disciplines which are closely aligned with it are planetary geology, space physics, cosmochemistry, etc. The chapter closely examines these fields of planetary science to provide an extensive understanding of the subject.

Planetary science is the scientific study of planets and their planetary systems which includes moons, ring systems, gas clouds, and magnetospheres. It involves understanding how planetary systems formed, how these systems work and how all their components interact. It is a cross-discipline field including aspects of astronomy, atmospheric science, geology, space physics, biology and chemistry.

What are the big scientific questions that Planetary Science tries to answer?

- How did the Solar System form?
- What are the conditions for life?
- How does the Solar System work?
- How does the Sun influence planets and planetary systems?
- How do planetary systems evolve?
- How are these questions answered?

Planetary scientists use lots of different techniques involving lots of branches of science to try to answer these questions:

- Use unmanned robot spacecraft to make measurements of the planets and their planetary systems.

- Use telescopes on the ground or in space to make observations of distant planets, moons, etc.
- Carry out experimental work in laboratories or in at different sites on the Earth which replicate different planetary environments.
- Do theoretical work involving mathematical work and building models of planetary systems or their components using computers.

Example: The Discovery of Large Organic Molecules in Titan's Atmosphere

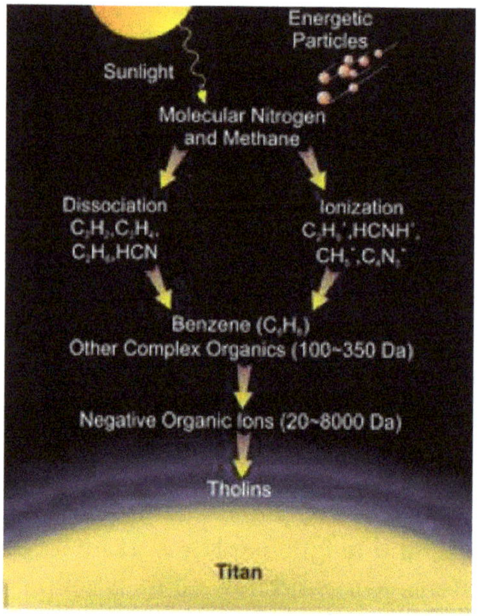

Tholin formation in Titan's upper atmosphere

Titan is Saturn's largest moon and the second largest moon in the Solar System. MSSL Planetary scientists used measurements made by the Cassini spacecraft, as it flew through Titan's upper atmosphere, to discover very large electrically charged organic molecules in Titan's atmosphere. By measuring their mass and considering the chemistry involve to create them we found out that they could be created by complex chemistry in the atmosphere. The energy to drive this process comes from sunlight and from particles in Saturn's magnetosphere raining down on the atmosphere. These particles slowly rain down onto Titan's surface, becoming neutral as they do so, and become a type of heavy molecule known as a Tholin. We think these are important for the origins of life and may have been important early in the history of the Earth. So in a sense Titan is like a window into the past history of the Earth.

Why are these Questions Important?

Where did the Solar System come from? Where did life come from? These sorts of questions are some of the most profound we can ask. It is also important to study many planetary systems as that tells us more about our own planet Earth and helps us understand effects such as space weather and climate change.

Planetary Geology

Planetary geology is a branch of geology devoted to the study of structure, composition, processes, and origin of major and minor planetary bodies in our solar system and beyond, and to the effects of interaction between planetary bodies within our solar system. Planetary geology interfaces with many other fields including astronomy, biology, chemistry, and physics. Planetary geologists work in the field, laboratory, and—indirectly—in outer space through imagery and other data returned by spacecraft. One planetary geologist has visited another planetary body (Dr. Harrison Schmitt worked on the Moon during the Apollo 17 mission in 1974) and perhaps someday others will visit the Moon, Mars, and other planetary bodies.

The goal of planetary geology studies of other planetary bodies is to understand the origin of the features seen (and samples returned, if such are available). Planetary geology studies usually relate to the quest for an understanding of the geological history of the body from its formation during accretion from the early solar nebula to its present condition. Most planetary bodies in the solar system of any significant size have passed through stages of accretion, internal heating and differentiation, and surficial and tectonic evolution. However, some objects have not, for example, small asteroids, comets, and smaller solar system debris such as meteorites. These objects can give important clues to the nature of the origin of our solar system, as they contain unaltered primordial material from the original nebular dust cloud. An important sub-discipline within planetary geology is meteoritics, the study of meteorites.

Most initial planetary geology studies help establish a temporal (time) framework for major events in the history of a planetary body. This framework, called a geologic time scale, is used to help fit events into a sequential order. On Earth, this kind of framework was established over a period of about 200 years using fossils, age relations among rocks, and absolute (radiometric) age dating. For planetary bodies like the Moon and Mars, for which we have some surface samples (note—there are a few Martian and Lunar meteorites plus the samples returned by Lunar landing missions), some radiometric age dating is possible. Mainly, planetary age relations are established by examining photographic evidence such as impact-crater densities, cross-cutting relationships, and superposition relationships (layers lying on top of one another). This is worked out by careful geologic mapping using photographic mosaics of planetary surfaces.

Planetary geologists are also interested in studying interactions among planetary bodies. When one planetary body impacts another, great energy can be released. Large body impacts represent the most energetic geologic process known. Very large body impacts can release the energy equivalent to thousands or even millions of megatons of TNT (megaton = 1 million tons) in a few seconds. Impact events produce impact craters, whose size and shape are controlled by factors like planetary gravity, strength of crustal materials, and density of planetary atmosphere. The impact crater is the most common feature on rocky and icy planetary surfaces in the solar system, except where "resurfacing" processes (like volcanism, erosion and sedimentation, and crustal melting) remove impact craters faster than they can form.

Studies of large impact craters on Earth have lead geologists to new understandings of how large body impacts, and the ecological disasters that they can cause, have affected evolution of life on Earth. Some large body impact events have been strongly connected to mass extinctions of life on

Earth, a driving force in evolutionary change. The best modern example of this is the great Cretaceous-Tertiary mass extinction event (65 million years ago), which has been dated as the same age as two large impact craters on Earth (Chicxulub crater in Mexico, nearly 112 mi [180 km] in diameter, and Boltysh crater in the Ukraine, about 15 mi [24 km] in diameter). Planetary geologists are currently working on theories about possible comet "swarms" due to perturbations in the Ort cloud or other disruptive mechanisms which may have made time intervals in the past more dangerous for life on Earth due to cosmic impacts.

Studies of impact craters hold considerable academic interest for investigators, but on the practical side, considerable natural resources have been discovered within impact craters. For this reason, their study has been supported by petroleum, mining, and groundwater concerns. The intensive fracturing caused by impact creates avenues for fluid movement within the affected crustal area, causing entrapment of hydrocarbons and emplacement of some types of fluid-borne mineral resources. Groundwater accessibility and quality is either enhanced or degraded in impact areas, depending upon geological conditions.

Studies of impacts of large comets may have other implications for solar system evolution, as recent theories suggest such impacts may have delivered significant amounts of water and primitive organic material to Earth and Mars during a very early phase of their evolution. Recent investigations of the origin of Earth's moon suggest that a major impact event that penetrated Earth's crust ejected material into orbit that eventually became the Moon. This theory helps explain some of the unusual chemical characteristics of lunar rocks and their similarity to Earth's upper mantle.

In the future, planetary geology will expand to the study of extra-solar system bodies. This future area for planetary geology investigation will help shed more light on the origin of our own solar system and on solar system formation in this part of our galaxy.

Cosmochemistry

Cosmochemistry is the science of the chemical composition of cosmic bodies, the laws of abundance and distribution of chemical elements throughout the universe, and the processes of the combination and migration of atoms during the formation of cosmic matter. Geochemistry is the best studied branch of cosmochemistry.

Cosmochemistry primarily investigates the "cold" processes on the atomic-molecular level of the interaction of substances, while physics primarily studies the "hot" nuclear processes in space, for example, the plasma state of matter and the formation of chemical elements within stars. Cosmochemistry, a new field of knowledge, has developed rapidly since the second half of the 20th century owing to the advancements in space exploration. Early investigations of chemical processes in space and of the composition of cosmic bodies were primarily conducted using spectral analysis of solar and stellar radiation and, on occasion, through the spectral analysis of the outer atmospheric layers of planets. This method led to the discovery of the element helium on the sun even before it was found to exist on the earth. The only direct method of studying cosmic bodies was the analysis of the chemical and phase compositions of various meteorites that had fallen onto the earth.

In this way, considerable data were accumulated having fundamental significance for further development of cosmochemistry. The advances in astronautics, the flights of unmanned space probes to the moon and the planets (Mars, Venus), and, finally, man's landing on the moon have created new possibilities in Cosmochemistry. These include the direct study of lunar rock samples by astronauts or the gathering of soil samples with automatic (mobile or stationary) devices and their return to the earth for further study in chemistry laboratories. Furthermore, automatic landing vehicles have made it possible to study matter and the conditions determining its presence in the atmosphere and on the surface of other planets, Mars and Venus in particular.

One of the most important tasks of Cosmochemistry is to study the evolution of cosmic bodies on the basis of the composition and abundance of chemical elements and to explain the origin and history of these bodies from a chemical point of view. Attention is focused on problems involving the abundance and distribution of chemical elements. The abundance of chemical elements in space is determined by their formation in the interior of stars. The chemical composition of the sun, of the terrestrial planets of the solar system, and of meteorites appears to be almost identical. The formation of nuclei of chemical elements is connected with the various nuclear processes within the stars. Therefore, at various stages in their evolution different stars and star systems do not have identical chemical compositions. Stars with particularly strong spectral lines of Ba, Mg, Li, or other elements are known to exist.

The distribution of chemical elements in different phases in cosmic processes is highly varied. The following factors affect in various ways the state of aggregation and state of phase of a substance in space at various stages of its transformation: (1) the vast temperature range, from stellar to absolute zero; (2) the vast range of pressures, from millions of atmospheres under planetary and stellar conditions to a cosmic vacuum; (3) highly penetrating galactic and solar radiation of varying composition and intensity; (4) radiation accompanying the transformation of unstable atoms into stable ones; and (5) magnetic, gravitational, and other physical fields. It has been established that all these factors affect the composition of the material in the planets' inner crust, of the planets' gaseous envelopes, of the meteorite material, and of cosmic dust. In this case, the fractionation of matter in space affects both the atomic and isotopic compositions. The determination of isotopic equilibriums, which arise in the presence of radiation, makes it possible to investigate thoroughly the history of formation of planets, asteroids, and meteorites and to establish the age of the processes.

Owing to the extreme conditions in space, there exist processes and states of matter that are not encountered on the earth. These include (1) the plasma state of stellar material (for example, the sun); (2) the condensation of He, H2, CH4, NH3, and other highly volatile gases in the atmospheres of the planets at very low temperatures; (3) the formation of noncorrosive iron in a cosmic vacuum during explosions on the moon; (4) the chondritic structure of matter in stony meteorites; and (5) the formation of complex organic matter in meteorites and, probably, on planet surfaces (for example, Mars). Small concentrations of atoms and molecules of many elements, as well as minerals (quartz, silicates, graphite), are found to exist in interstellar space. Further-more, the synthesis of various complex organic compounds (arising from primary solar gases H, CO, NH3, O2, N2, S, and other simple compounds under equilibrium conditions in the presence of radiation) takes place. All these organic substances in meteorites and interstellar space are optically inactive.

With the development of astrophysics and a number of other sciences, the possibilities of obtaining information relating to cosmochemistry were greatly expanded. For example, radio astronomy is employed to search for molecules in the interstellar medium. More than 20 types of molecules had been discovered in interstellar space by the end of 1972, including several rather complex organic molecules containing up to seven atoms each. It was established that their observed concentrations were 10 to 100 million times smaller than the concentration of hydrogen. Radio astronomy also permits the investigation of the isotopic composition of interstellar gas by comparing the microwave lines of the isotopic varieties of one molecule (for example, $H_2{}^{12}CO$ and $H_2{}^{13}CO$) and the verification of the validity of existing theories of the origin of chemical elements.

Of great importance in the study of the chemistry of space is the investigation of the complex multistage condensation of matter in the low-temperature plasma, for example, the transition of solar material into the solids of the planets, asteroids, and meteorites, accompanied by condensation growth, accretion (increase in mass, the "growth" of any substance by the external addition of particles, for example, from a gas-dust cloud), and the agglomeration of primary aggregates (phases) accompanied by loss of volatiles into the cosmic vacuum. In a cosmic vacuum, at relatively low temperatures (5000°-10,000 °C), solid phases of varying chemical composition (depending on temperature), characterized by different bonding energies, oxidation potentials, and other factors, are successively precipitated from the cooling plasma. For example, silicate, metallic, sulfide, chromite, phosphide, carbide, and other phases are differentiated in chondrites; these phases agglomerate at some point in their history into a stony meteorite and, probably, in the same manner into the matter making up the terrestrial planets.

Furthermore, the planets undergo a differentiation of solid, cooled matter into shells—metallic core, silicate phases (mantle and crust), and atmosphere. This differentiation is the result of the secondary warming up of the planetary material by radiogenic heat emitted during the decay of radioactive isotopes of potassium, uranium, thorium, and possibly other elements. Such a process of melting and degassing of matter during volcanism is characteristic of the moon, the earth, Mars, and Venus. It is based on the universal principle of zone melting, which separates the low-melting material (for example, the crust and atmosphere) from the high-melting material of planetary mantles. For example, primary solar matter has the ratio $Si/Mg \approx 1$, while for the planetary crust produced by melting of the mantle the ratio is $Si/Mg \approx 6.5$. The preservation and nature of outer planetary shells depend primarily on the mass of the planets and the distances of the planets from the sun (for example, the lowdensity atmosphere of Mars and the high-density atmosphere of Venus). Owing to the proximity of Venus to the sun, a "greenhouse" effect has resulted from the CO_2 in the Venusian atmosphere: at temperatures higher than 300 °C, the process $CaCO_3 + SiO_2 \rightarrow CaSiO_3 + CO_2$ reaches a state of equilibrium during which a 97 percent CO_2 content at a pressure of 90 atm is observed. Using the moon as an example, secondary (volcanic) gases cannot be retained by a celestial body of small mass.

Collisions in cosmic space (either among particles of meteoritic material or during the fall of meteorites and other particles onto planet surfaces) caused by enormous velocities can create a thermal explosion, which leaves traces in the structure of solid cosmic bodies and forms meteorite craters. Exchange of matter occurs between cosmic bodies. For example, according to a minimum estimate no less than 1 X 104 tons of cosmic dust whose composition is known is precipitated onto the earth annually. Among the stony meteorites that fall to the earth are the basalt achondrites, which are similar in composition to lunar rocks and the basalt on the earth's surface ($Si/Mg \approx 6.5$). In view of this, it has been hypothesized that the source of these achondrites is the moon (surface rocks of its crust).

These and other processes in space are accompanied by the irradiation of matter (high-energy galactic and solar radiation) during the numerous stages of its transformation, and this leads, in particular, to the transformation of one type of isotope into another and, in the general case, to a change in the isotopic or atomic composition. The more prolonged and varied the processes, the further is its chemical composition from the primary stellar (solar) composition. At the same time, the isotopic com-position of cosmic matter (for example, meteorites) makes it possible to determine the composition, intensity, and modulation of past galactic radiation.

Space Physics

Space physics is the study of plasmas as they occur naturally in the Earth's upper atmosphere (aeronomy) and within the Solar System. As such, it encompasses a far-ranging number of topics, such as heliophysics which includes the solar physics of the Sun: the solar wind, planetary magnetospheres and ionospheres, auroras, cosmic rays, and synchrotron radiation. Space physics is a fundamental part of the study of space weather and has important implications in not only to understanding the universe, but also for practical everyday life, including the operations of communications and weather satellites.

Space physics is distinct from astrophysical plasma and the field of astrophysics, which studies similar plasma phenomena beyond the Solar System. Space physics utilizes in situ measurements from high altitude rockets and spacecraft, in contrast to astrophysical plasma that relies deduction of theory and astronomical observation.

Space physics can be traced to the Chinese who discovered the principle of the compass, but did not understand how it worked. During the 16th century, in De Magnete, William Gilbert gave the first description of the Earth's magnetic field, showing that the Earth itself is a great magnet, which explained why a compass needle points north. Deviations of the compass needle magnetic declination were recorded on navigation charts, and a detailed study of the declination near London by watchmaker George Graham resulted in the discovery of irregular magnetic fluctuations that we now call magnetic storms, so named by Alexander Von Humboldt. Gauss and William Weber made very careful measurements of Earth's magnetic field which showed systematic variations and random fluctuations. This suggested that the Earth was not an isolated body, but was influenced by external forces – especially from the Sun and the appearance of sunspots. A relationship between individual aurora and accompanying geomagnetic disturbances was noticed by Anders Celsius and Olof Peter Hiorter in 1747. In 1860, Elias Loomis (1811–1889) showed that the highest incidence of aurora is seen inside an oval of 20 - 25 degrees around the magnetic pole. In 1881, Hermann Fritz published a map of the "isochasms" or lines of constant magnetic field.

In the late 1870s, Henri Becquerel offered the first physical explanation for the statistical correlations that had been recorded: sunspots must be a source of fast protons. They are guided to the poles by the Earth's magnetic field. In the early twentieth century, these ideas led Kristian Birkeland to build a terella, or laboratory device which simulates the Earth's magnetic field in a vacuum chamber, and which uses a cathode ray tube to simulate the energetic particles which compose the solar wind. A theory began to be formulated about the interaction between the Earth's magnetic field and the solar wind.

Space physics did not begin in earnest, however, until the first in situ measurements in the early 1950s, when a team led by Van Allen launched the first rockets to a height around 110 km. In 1958, a Geiger counter on board the first US satellite, Explorer 1 detected the Earth's radiation belts, later named the Van Allen belts. The boundary between the Earth's magnetic field and interplanetary space was studied by Explorer 10. Future space craft would travel outside Earth orbit and study the composition and structure of the solar wind in much greater detail. These include WIND (spacecraft), (1994), Advanced Composition Explorer (ACE), Ulysses, the Interstellar Boundary Explorer (IBEX) in 2008, and Parker Solar Probe. Other spacecraft would study the sun, such as STEREO and Solar and Heliospheric Observatory (SOHO).

Solar System

Solar system is an assemblage consisting of the Sun—an average star in the Milky Way Galaxy—and those bodies orbiting around it: 8 (formerly 9) planets with about 170 known planetary satellites (moons); countless asteroids, some with their own satellites; comets and other icy bodies; and vast reaches of highly tenuous gas and dust known as the interplanetary medium.

The eight planets of the solar system and Pluto, in a montage of images scaled to show the approximate sizes of the bodies relative to one another. Outward from the Sun, which is represented to scale by the yellow segment at the extreme left, are the four rocky terrestrial planets (Mercury, Venus, Earth, and Mars), the four hydrogen-rich giant planets (Jupiter, Saturn, Uranus, and Neptune), and icy, comparatively tiny Pluto.

The Sun, Moon, and brightest planets were visible to the naked eyes of ancient astronomers, and their observations and calculations of the movements of these bodies gave rise to the science of astronomy. Today the amount of information on the motions, properties, and compositions of the planets and smaller bodies has grown to immense proportions, and the range of observational instruments has extended far beyond the solar system to other galaxies and the edge of the known universe. Yet the solar system and its immediate outer boundary still represent the limit of our physical reach, and they remain the core of our theoretical understanding of the cosmos as well. Earth-launched space probes and landers have gathered data on planets, moons, asteroids, and other bodies, and this data has been added to the measurements collected with telescopes and other instruments from below and above Earth's atmosphere and to the information extracted from meteorites and from Moon rocks returned by astronauts. All this information is scrutinized in attempts to understand in detail the origin and evolution of the solar system—a goal toward which astronomers continue to make great strides.

Composition of the Solar System

Located at the centre of the solar system and influencing the motion of all the other bodies through its gravitational force is the Sun, which in itself contains more than 99 percent of the mass of the system. The planets, in order of their distance outward from the Sun, are Mercury, Venus, Earth, Mars, Jupiter, Saturn, Uranus, and Neptune. Four planets—Jupiter through Neptune—have ring systems, and all but Mercury and Venus have one or more moons. Pluto had been officially listed among the planets since it was discovered in 1930 orbiting beyond Neptune, but in 1992 an icy object was discovered still farther from the Sun than Pluto. Many other such discoveries followed, including an object named Eris that appears to be at least as large as Pluto. It became apparent that Pluto was simply one of the larger members of this new group of objects, collectively known as the Kuiper belt. Accordingly, in August 2006 the International Astronomical Union (IAU), the organization charged by the scientific community with classifying astronomical objects, voted to revoke Pluto's planetary status and place it under a new classification called dwarf planet.

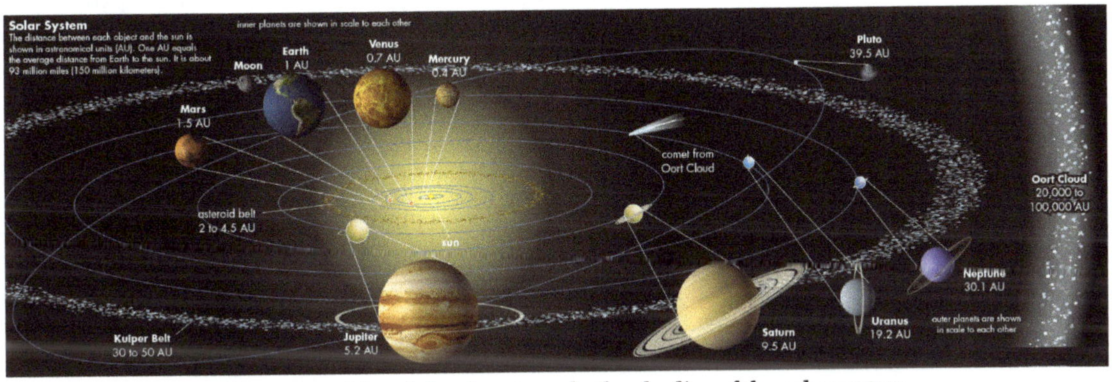

Orbits: The orbits of the planets and other bodies of the solar system.

Any natural solar system object other than the Sun, a planet, a dwarf planet, or a moon is called a small body; these include asteroids, meteoroids, and comets. Most of the several hundred thousand asteroids, or minor planets, orbit between Mars and Jupiter in a nearly flat ring called the asteroid belt. The myriad fragments of asteroids and other small pieces of solid matter (smaller than a few tens of metres across) that populate interplanetary space are often termed meteoroids to distinguish them from the larger asteroidal bodies.

The solar system's several billion comets are found mainly in two distinct reservoirs. The more-distant one, called the Oort cloud, is a spherical shell surrounding the solar system at a distance of approximately 50,000 astronomical units (AU)—more than 1,000 times the distance of Pluto's orbit. The other reservoir, the Kuiper belt, is a thick disk-shaped zone whose main concentration extends 30–50 AU from the Sun, beyond the orbit of Neptune but including a portion of the orbit of Pluto. (One astronomical unit is the average distance from Earth to the Sun—about 150 million km [93 million miles].) Just as asteroids can be regarded as rocky debris left over from the formation of the inner planets, Pluto, its moon Charon, Eris, and the myriad other Kuiper belt objects can be seen as surviving representatives of the icy bodies that accreted to form the cores of Neptune and Uranus. As such, Pluto and Charon may also be considered to be very large comet nuclei. The Centaur objects, a population of comet nuclei having diameters as large as 200 km (125 miles), orbit the Sun between Jupiter and Neptune, probably having been gravitationally perturbed inward from the Kuiper belt. The interplanetary medium—an exceedingly tenuous plasma (ionized gas) laced with concentrations of dust particles—extends outward from the Sun to about 123 AU.

Orbits

All the planets and dwarf planets, the rocky asteroids, and the icy bodies in the Kuiper belt move around the Sun in elliptical orbits in the same direction that the Sun rotates. This motion is termed prograde, or direct, motion. Looking down on the system from a vantage point above Earth's North Pole, an observer would find that all these orbital motions are in a counterclockwise direction. In striking contrast, the comet nuclei in the Oort cloud are in orbits having random directions, corresponding to their spherical distribution around the plane of the planets.

The shape of an object's orbit is defined in terms of its eccentricity. For a perfectly circular orbit, the eccentricity is 0; with increasing elongation of the orbit's shape, the eccentricity increases toward a value of 1, the eccentricity of a parabola. Of the eight major planets, Venus and Neptune have the most circular orbits around the Sun, with eccentricities of 0.007 and 0.009, respectively. Mercury, the closest planet, has the highest eccentricity, with 0.21; the dwarf planet Pluto, with 0.25, is even more eccentric. Another defining attribute of an object's orbit around the Sun is its inclination, which is the angle that it makes with the plane of Earth's orbit—the ecliptic plane. Again, of the planets, Mercury's has the greatest inclination, its orbit lying at 7° to the ecliptic; Pluto's orbit, by comparison, is much more steeply inclined, at 17.1°. The orbits of the small bodies generally have both higher eccentricities and higher inclinations than those of the planets. Some comets from the Oort cloud have inclinations greater than 90°; their motion around the Sun is thus opposite that of the Sun's rotation, or retrograde.

Planets and their Moons

The eight planets can be divided into two distinct categories on the basis of their densities (mass per unit volume). The four inner, or terrestrial, planets—Mercury, Venus, Earth, and Mars—have rocky compositions and densities greater than 3 grams per cubic cm. (Water has a density of 1 gram per cubic cm.) In contrast, the four outer planets, also called the Jovian, or giant, planets—Jupiter, Saturn, Uranus, and Neptune—are large objects with densities less than 2 grams per cubic cm; they are composed primarily of hydrogen and helium (Jupiter and Saturn) or of ice, rock, hydrogen, and helium (Uranus and Neptune). The dwarf planet Pluto is unique—an icy, low-density body smaller than Earth's Moon, more similar to comets or to the large icy moons of the outer planets than to any of the planets themselves. Its acceptance as a member of the Kuiper belt explains these anomalies.

The relatively small inner planets have solid surfaces, lack ring systems, and have few or no moons. The atmospheres of Venus, Earth, and Mars are composed of a significant percentage of oxidized compounds such as carbon dioxide. Among the inner planets, only Earth has a strong magnetic field, which shields it from the interplanetary medium. The magnetic field traps some of the electrically charged particles of the interplanetary medium inside a region around Earth known as the magnetosphere. Heavy concentrations of these high-energy particles occur in the Van Allen belts in the inner part of the magnetosphere.

The four giant outer planets are much more massive than the terrestrial planets and have immense atmospheres composed mainly of hydrogen and helium. They have no solid surfaces, however, and their densities are so low that one of them, Saturn, would actually float in water. Each of the outer planets has a magnetic field, a ring system, and many known moons, with more likely to be discovered. Pluto has no known rings and only five known moons. Several other Kuiper belt objects and some asteroids also have moons of their own.

Jovian planets: The Jovian—or gaseous, Jupiter-like—planets.

Most of the known moons move around their respective planets in the same direction that the planets orbit the Sun. They are extremely diverse, representing a wide range of environments. Jupiter is orbited by Io, a body wracked by intense volcanism, while Saturn's largest moon, Titan—a body larger than the terrestrial planet Mercury—exhibits a primitive atmosphere denser than that of Earth. Triton moves in a retrograde orbit around Neptune—that is, opposite to the direction of the planet's orbit around the Sun—and features plumes of material rising through its tenuous atmosphere from a surface whose temperature is only 37 kelvins (K; −393 °F, −236 °C).

Asteroids and Comets

The asteroids and comets are remnants of the planet-building process in the inner and outer solar system, respectively. The asteroid belt is home to rocky bodies ranging in size from the largest known asteroid, Ceres (also classified by the IAU as a dwarf planet), with a diameter of roughly 940 km (585 miles), to microscopic dust particles that are dispersed throughout the belt. Some asteroids travel in paths that cross the orbit of Earth, providing opportunities for collisions with the planet. The rare collisions of relatively large objects (those with diameters greater than about 1 km [0.6 mile]) with Earth can be devastating, as in the case of the asteroid impact that is thought to have been responsible for the massive extinction of species at the end of the Cretaceous Period 65 million years ago. More commonly, the impacting objects are much smaller, reaching Earth's surface as meteorites. Asteroid observations from Earth, which have been confirmed by spacecraft flybys, indicate that some asteroids are mainly metal (principally iron), others are stony, and still others are rich in organic compounds, resembling the carbonaceous chondrite meteorites. The asteroids that have been visited by spacecraft are irregularly shaped objects pockmarked with craters; some of them have retained very primitive material from the early days of the solar system.

Gaspra, an asteroid of the main belt, in a composite of two images taken by the Galileo spacecraft during its flyby on October 29, 1991. Pocked with numerous small craters, Gaspra measures about 20 km (12 miles) in its longest dimension. Its irregular shape and groovelike linear markings suggest that it was once part of a larger body that experienced one or more shattering collisions. Colours in the composite image have been enhanced by computer to highlight subtle variations in reflectivity and other surface characteristics.

The physical characteristics of comet nuclei are fundamentally different from those of asteroids. Ices are their main constituent, predominantly frozen water, but frozen carbon dioxide, carbon monoxide, methanol, and other ices are also present. These cosmic ice balls are laced with rock dust and a rich variety of organic compounds, many of which are collected in tiny grains. Some comets may have more such "dirt" than ice.

Comets can be classified according to their orbital period, the time it takes for them to revolve around the Sun. Comets that have orbital periods greater than 200 years (and usually much greater) are called long-period comets; those that make a return appearance in less time are short-period comets. Each kind appears to have a distinct source.

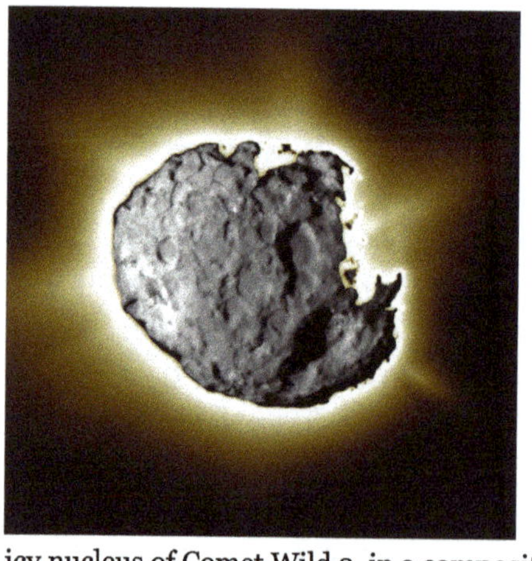

Comet Wild 2 The icy nucleus of Comet Wild 2, in a composite image taken by the U.S. Stardust spacecraft during its close approach to the comet on January 2, 2004.

The image consists of a short exposure to resolve fine details of the surface and a longer exposure to capture the faint jets of gas and dust streaming into space. Wild 2's nucleus is about 5 km (3 miles) across.

The nucleus of a typical long-period comet is irregularly shaped and a few kilometres across. It can have an orbital period of millions of years, and it spends most of its life at immense distances from the Sun, as much as one-fifth of the way to the nearest star. This is the realm of the Oort cloud. The comet nuclei in this spherical shell are too distant to be visible from Earth. The presence of the cloud is presumed from the highly elliptical orbits—with eccentricities close to 1—in which the long-period comets are observed as they approach and then swing around the Sun. Their orbits can be inclined in any direction—hence the inference that the Oort cloud is spherical. In contrast, most short-period comets, particularly those with periods of 20 years or less, move in rounder, prograde orbits near the plane of the solar system. Their source is believed to be the much nearer Kuiper belt, which lies in the plane of the solar system beyond the orbit of Neptune. Comet nuclei in the Kuiper belt have been photographed from Earth with large telescopes.

As comet nuclei trace out the parts of their orbits closest to the Sun, they are warmed through solar heating and begin to shed gases and dust, which form the familiar fuzzy-looking comas and long, wispy tails. The gas dissipates into space, but the grains of silicates and organic compounds remain to orbit the Sun along paths very similar to that of the parent comet. When Earth's path around the Sun intersects one of these dust-populated orbits, a meteor shower occurs. During such an event, nighttime observers may see tens to hundreds of so-called shooting stars per hour as the dust grains burn up in the upper atmosphere of Earth. Although many random meteors can be observed nightly, they occur at a much higher rate during a meteor shower. Even on an average day, Earth's atmosphere is bombarded with more than 80 tons of dust grains, mostly asteroidal and cometary debris.

In addition to particles of debris, the space through which the planets travel contains protons, electrons, and ions of the abundant elements, all streaming outward from the Sun in the form of the solar wind. Occasional giant solar flares, short-lived eruptions on the Sun's surface, expel matter (along with high-energy radiation) that contributes to this interplanetary medium.

In 2012 the space probe Voyager 1 crossed the boundary between the interplanetary medium and the interstellar medium—a region called the heliopause. Since passing through the heliopause, Voyager 1 has been able to measure the properties of interstellar space.

Star System

A star system or stellar system is a small number of stars that orbit each other, bound by gravitational attraction. A large number of stars bound by gravitation is generally called a star cluster or galaxy, although, broadly speaking, they are also star systems. Star systems are not to be confused with planetary systems, which include planets and similar bodies such as comets.

A star system of two stars is known as a binary star, binary star system or physical double star. If there are no tidal effects, no perturbation from other forces, and no transfer of mass from one star to the other, such a system is stable, and both stars will trace out an elliptic orbit around the barycenter of the system indefinitely. Examples of binary systems are Sirius, Procyon and Cygnus X-1, the last of which probably consists of a star and a black hole.

A multiple star system consists of three or more stars that appear from Earth to be close to one another in the sky. This may result from the stars actually being physically close and gravitationally bound to each other, in which case it is a physical multiple star, or this closeness may be merely apparent, in which case it is an optical multiple star (meaning that the stars may appear to be close to each other when viewed from planet Earth, as they both seem to occupy the same point in the sky, but in reality, one star may be much further away from Earth than the other, which is not readily apparent unless one can view them from a different angle). Physical multiple stars are also commonly called multiple stars or multiple star systems.

Most multiple star systems are triple stars. Systems with four or more components are less likely to occur. Multiple-star systems are called triple, trinary or ternary if they contain three stars; quadruple or quaternary if they contain four stars; quintuple or quintenary with five stars; sextuple or sextenary with six stars; septuple or septenary with seven stars. These systems are smaller than open star clusters, which have more complex dynamics and typically have from 100 to 1,000 stars. Most multiple star systems known are triple; for higher multiplicities, the number of known systems with a given multiplicity decreases exponentially with multiplicity. For example, in the 1999 revision of Tokovinin's catalog of physical multiple stars, 551 out of the 728 systems described are triple. However, because of selection effects, knowledge of these statistics is very incomplete.

Multiple-star systems can be divided into two main dynamical classes: hierarchical systems which are stable and consist of nested orbits that don't interact much and so each level of the hierarchy can be treated as a Two-body problem, or the trapezia which have unstable strongly interacting orbits and are modelled as an n-body problem, exhibiting chaotic behavior.

Hierarchical Systems

Star system named DI Cha. While only two stars are apparent, it is actually a quadruple system containing two sets of binary stars.

Most multiple-star systems are organized in what is called a hierarchical system: the stars in the system can be divided into two smaller groups, each of which traverses a larger orbit around the system's center of mass. Each of these smaller groups must also be hierarchical, which means that they must be divided into smaller subgroups which themselves are hierarchical, and so on. Each level of the hierarchy can be treated as a two-body problem by considering close pairs as if they were a single star. In these systems there is little interaction between the orbits and the stars' motion will continue to approximate stable Keplerian orbits around the system's center of mass, unlike the unstable trapezia systems or the even more complex dynamics of the large number of stars in star clusters and galaxies.

Triple Star Systems

In a physical triple star system, each star orbits the center of mass of the system. Usually, two of the stars form a close binary system, and the third orbits this pair at a distance much larger than that of the binary orbit. This arrangement is called hierarchical. The reason for this is that if the inner and outer orbits are comparable in size, the system may become dynamically unstable, leading to a star being ejected from the system. Triple stars that are not all gravitationally bound might comprise a physical binary and an optical companion, such as Beta Cephei, or rarely, a purely optical triple star, such as Gamma Serpentis.

Higher Multiplicities

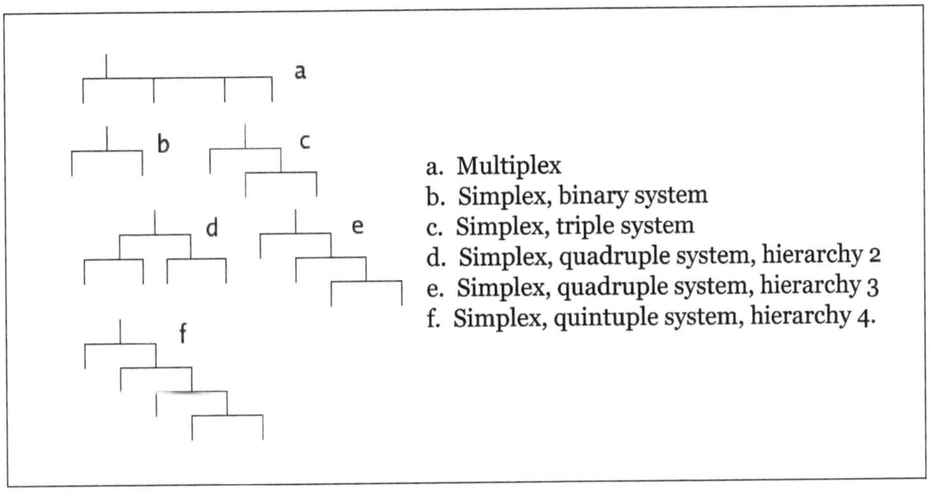

Mobile diagrams.

Hierarchical multiple star systems with more than three stars can produce a number of more complicated arrangements, which can be illustrated by what Evans (1968) has called a *mobile diagram*. These are similar to ornamental mobiles hung from the ceiling. Some examples can be seen in the figure to the right. Each level of the diagram illustrates the decomposition of the system into two or more systems with smaller size. Evans calls a diagram *multiplex* if there is a node with more than two children, i.e. if the decomposition of some subsystem involves two or more orbits with comparable size. Because, as we have already seen for triple stars, this may be unstable, multiple stars are expected to be *simplex*, meaning that at each level there are exactly two children. Evans calls the number of levels in the diagram its *hierarchy*.

- A simplex diagram of hierarchy 1, as in (b), describes a binary system.
- A simplex diagram of hierarchy 2 may describe a triple system, as in (c), or a quadruple system, as in (d).
- A simplex diagram of hierarchy 3 may describe a system with anywhere from four to eight components. The mobile diagram in (e) shows an example of a quadruple system with hierarchy 3, consisting of a single distant component orbiting a close binary system, with one of the components of the close binary being an even closer binary.
- A real example of a system with hierarchy 3 is Castor, also known as Alpha Geminorum or

α Gem. It consists of what appears to be a visual binary star which, upon closer inspection, can be seen to consist of two spectroscopic binary stars. By itself, this would be a quadruple hierarchy 2 system as in (d), but it is orbited by a fainter more distant component, which is also a close red dwarf binary. This forms a sextuple system of hierarchy 3.

- The maximum hierarchy occurring in A. A. Tokovinin's Multiple Star Catalogue, as of 1999, is 4. For example, the stars Gliese 644A and Gliese 644B form what appears to be a close visual binary star; because Gliese 644B is a spectroscopic binary, this is actually a triple system. The triple system has the more distant visual companion Gliese 643 and the still more distant visual companion Gliese 644C, which, because of their common motion with Gliese 644AB, are thought to be gravitationally bound to the triple system. This forms a quintuple system whose mobile diagram would be the diagram of level 4 appearing in (f).

Higher hierarchies are also possible. Most of these higher hierarchies either are stable or suffer from internal perturbations. Others consider complex multiple stars will in time theoretically disintegrate into less complex multiple stars, like more common observed triples or quadruples are possible.

Trapezia

Trapezia are usually very young, unstable systems. These are thought to form in stellar nurseries, and quickly fragment into stable multiple stars, which in the process may eject components as galactic high-velocity stars. They are named after the multiple star system known as the Trapezium Cluster in the heart of the Orion Nebula. Such systems are not rare, and commonly appear close to or within bright nebulae. These stars have no standard hierarchical arrangements, but compete for stable orbits. This relationship is called *interplay*. Such stars eventually settle down to a close binary with a distant companion, with the other star(s) previously in the system ejected into interstellar space at high velocities. Example of such events may explain the runaway stars that might have been ejected during a collision of two binary star groups or a multiple system. This event is credited with ejecting AE Aurigae, Mu Columbae and 53 Arietis at above 200 $km{\cdot}s^{-1}$ and has been traced to the Trapezium cluster in the Orion Nebula some two million years ago.

Designations and Nomenclature

Multiple Star Designations

The components of multiple stars can be specified by appending the suffixes *A*, *B*, *C*, etc., to the system's designation. Suffixes such as *AB* may be used to denote the pair consisting of *A* and *B*. The sequence of letters *B*, *C*, etc. may be assigned in order of separation from the component *A*. Components discovered close to an already known component may be assigned suffixes such as *Aa*, *Ba*, and so forth.

Nomenclature in the Multiple Star Catalogue

A.A. Tokovinin's Multiple Star Catalogue uses a system in which each subsystem in a mobile diagram is encoded by a sequence of digits. In the mobile diagram (d) above, for example, the widest system would be given the number 1, while the subsystem containing its primary component would be numbered 11 and the subsystem containing its secondary component would be numbered 12. Subsystems which would appear below this in the mobile diagram will be given numbers with three, four, or more digits. When describing a non-hierarchical system by this method, the same subsystem number will be used more than once; for example, a system with three visual components, A, B, and C, no two of which can be grouped into a subsystem, would have two subsystems numbered 1 denoting the two binaries AB and AC. In this case, if B and C were subsequently resolved into binaries, they would be given the subsystem numbers 12 and 13.

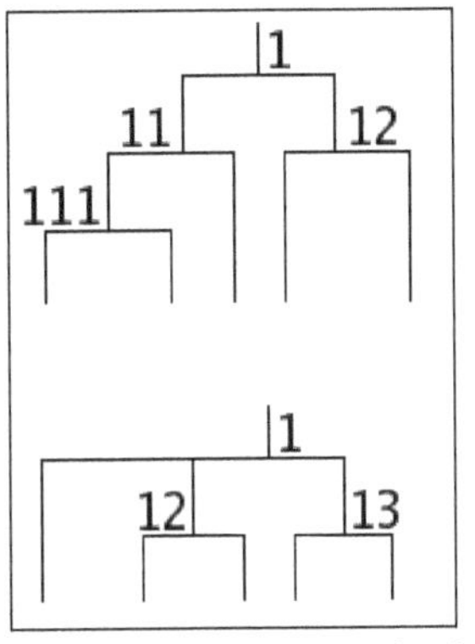

Subsystem notation in Tokovinin's Multiple Star Catalogue

Future Multiple Star System Nomenclature

The current nomenclature for double and multiple stars can cause confusion as binary stars discovered in different ways are given different designations (for example, discoverer designations for visual binary stars and variable star designations for eclipsing binary stars), and, worse, component letters may be assigned differently by different authors, so that, for example, one person's *A* can be another's *C*. Discussion starting in 1999 resulted in four proposed schemes to address this problem:

- KoMa, a hierarchical scheme using upper- and lower-case letters and Arabic and Roman numerals;
- The Urban/Corbin Designation Method, a hierarchical numeric scheme similar to the Dewey Decimal Classification system;
- The Sequential Designation Method, a non-hierarchical scheme in which components and subsystems are assigned numbers in order of discovery; and
- WMC, the Washington Multiplicity Catalog, a hierarchical scheme in which the suffixes used in the Washington Double Star Catalog are extended with additional suffixed letters and numbers.

For a designation system, identifying the hierarchy within the system has the advantage that it makes identifying subsystems and computing their properties easier. However, it causes problems when new components are discovered at a level above or intermediate to the existing hierarchy. In this case, part of the hierarchy will shift inwards. Components which are found to be nonexistent, or are later reassigned to a different subsystem, also cause problems.

During the 24th General Assembly of the International Astronomical Union in 2000, the WMC scheme was endorsed and it was resolved by Commissions 5, 8, 26, 42, and 45 that it should be expanded into a usable uniform designation scheme. A sample of a catalog using the WMC scheme, covering half an hour of right ascension, was later prepared. The issue was discussed again at the 25th General Assembly in 2003, and it was again resolved by commissions 5, 8, 26, 42, and 45, as well as the Working Group on Interferometry, that the WMC scheme should be expanded and further developed.

The sample WMC is hierarchically organized; the hierarchy used is based on observed orbital periods or separations. Since it contains many visual double stars, which may be optical rather than physical, this hierarchy may be only apparent. It uses upper-case letters (A, B) for the first level of the hierarchy, lower-case letters (a, b) for the second level, and numbers (1, 2) for the third. Subsequent levels would use alternating lower-case letters and numbers, but no examples of this were found in the sample.

References

- What-planetary-science, planetary-science, solar-system, research, mssl: ucl.ac.uk, Retrieved 19 April, 2019
- Planetary-geology, 5259, pages: jrank.org, Retrieved 25 February, 2019
- Kallenrode, may-britt (2004). Space physics: an introduction to plasmas and particles in the heliosphere and magnetospheres. Springer. Isbn 978-3-540-20617-0
- Solar-system, science: britannica.com, Retrieved 8 January, 2019
- P. 16, understanding variable stars, john r. Percy, cambridge: cambridge university press, 2007, isbn 0-521-23253-8
- Rubinov, a. V. (2004). “dynamical evolution of multiple stars: influence of the initial parameters of the system”. Astronomy reports. 48: 155–160. Bibcode:2004arep...48...45r. Doi:10.1134/1.1641122
- Cosmochemistry: thefreedictionary, Retrieved 6 July, 2019
- William i. Hartkopf & brian d. Mason. “addressing confusion in double star nomenclature: the washington multiplicity catalog”. United states naval observatory. Retrieved 12 september 2008

PERMISSIONS

All chapters in this book are published with permission under the Creative Commons Attribution Share Alike License or equivalent. Every chapter published in this book has been scrutinized by our experts. Their significance has been extensively debated. The topics covered herein carry significant information for a comprehensive understanding. They may even be implemented as practical applications or may be referred to as a beginning point for further studies.

We would like to thank the editorial team for lending their expertise to make the book truly unique. They have played a crucial role in the development of this book. Without their invaluable contributions this book wouldn't have been possible. They have made vital efforts to compile up to date information on the varied aspects of this subject to make this book a valuable addition to the collection of many professionals and students.

This book was conceptualized with the vision of imparting up-to-date and integrated information in this field. To ensure the same, a matchless editorial board was set up. Every individual on the board went through rigorous rounds of assessment to prove their worth. After which they invested a large part of their time researching and compiling the most relevant data for our readers.

The editorial board has been involved in producing this book since its inception. They have spent rigorous hours researching and exploring the diverse topics which have resulted in the successful publishing of this book. They have passed on their knowledge of decades through this book. To expedite this challenging task, the publisher supported the team at every step. A small team of assistant editors was also appointed to further simplify the editing procedure and attain best results for the readers.

Apart from the editorial board, the designing team has also invested a significant amount of their time in understanding the subject and creating the most relevant covers. They scrutinized every image to scout for the most suitable representation of the subject and create an appropriate cover for the book.

The publishing team has been an ardent support to the editorial, designing and production team. Their endless efforts to recruit the best for this project, has resulted in the accomplishment of this book. They are a veteran in the field of academics and their pool of knowledge is as vast as their experience in printing. Their expertise and guidance has proved useful at every step. Their uncompromising quality standards have made this book an exceptional effort. Their encouragement from time to time has been an inspiration for everyone.

The publisher and the editorial board hope that this book will prove to be a valuable piece of knowledge for students, practitioners and scholars across the globe.

INDEX

CPSIA information can be obtained
at www.ICGtesting.com
Printed in the USA
LVHW062011100122
708212LV00002B/172

9 781641 165709